A MANUAL OF
INTENSIONAL LOGIC

T0351385

A MANUAL OF INTENSIONAL LOGIC

Second Edition
Revised and Expanded

Johan van Benthem

CENTER FOR THE STUDY
OF LANGUAGE
AND INFORMATION

CSLI was founded early in 1983 by researchers from Stanford University, SRI International, and Xerox PARC to further research and development of integrated theories of language, information, and computation. CSLI headquarters and the publication offices are located at the Stanford site.

CSLI/SRI International
333 Ravenswood Avenue
Menlo Park, CA 94025

CSLI/Stanford
Ventura Hall
Stanford, CA 94305

CSLI/Xerox PARC
3333 Coyote Hill Road
Palo Alto, CA 94304

Printed in the United States

94 93 92 91 90 89 88   5 4 3 2 1

Library of Congress Cataloging-in-Publication Data

Benthem, J. F. A. K. van, 1949–
    A manual of intensional logic / Johan van Benthem. – – 2nd ed., rev. and expanded.
        p. cm. – – (CSLI lecture notes ; no. 1)
Includes bibliographies and index.
1. Logic. I. Stanford University. Center for the Study of Language and Information. II. Title. III. Series.
BC71.B38   1988
160—dc 19   87–34935
                                                         CIP

ISBN 0–937073–30–X
ISBN 0–937073–29–6 (pbk.)

To Marie Virgine

Contents

Acknowledgments

It is a pleasure to acknowledge some debts incurred when writing this manual, both the original and its revision.

The course upon which this book is based was taught at Stanford in the spring of 1984, during a stay made possible by the support of the Department of Philosophy, CSLI, and the Sloan Foundation. John Perry was especially instrumental in organizing this visit of an exponent (and product) of the "ancien régime" in intensional logic.

As for editorial assistance, and indeed guidance in the production of the work, the help of Dikran Karagueuzian and his collaborators has been invaluable. Their combination of efficiency and friendliness has deeply impressed this veteran of the Dutch academic bureaucracy.

Introduction

These notes form the content of a graduate course in Intensional Logic taught at Stanford University in the Winter of 1984.

Intensional Logic as understood here is a research program based upon the broad presupposition that so-called "intensional contexts" in natural language can be explained semantically by the idea of *multiple reference*. Thus, temporal contexts require shifting patterns of ordinary Tarski-style denotations through time, modal contexts are related to varying denotations in some set of relevant worlds, or situations. Instead of developing one most general framework for this technique, we shall consider the above, as well as other examples.

The inspiration for the semantic theories to be presented here has been diverse, coming from linguistics, traditional philosophy, but also, for example, from the philosophy of science. On the intellectual map of a logician there is nothing strange to this combination, these being his natural neighbors. We shall see these various forms of inspiration recur throughout our examples.

From a technical point of view, the multiple reference move has the advantage that classical notions and arguments often remain applicable. On the other hand, this also makes intensional logic, on the whole, an area of application rather than innovation for mainstream logic. Even so, quite a few technical developments have occurred in the area with intrinsic logical interest (witness R. A. Bull's reviews in the *Journal of Symbolic Logic*, vol. 47:2, 1982, p. 440–445, and vol. 48:2, 1983, p. 488–495). In these notes, no such topics are explored in any depth, our aim being rather a survey of trends and questions in the area.

One conspicuous feature of intensional semantics, setting it apart from its classical ancestor, is the role played by *patterns* of classical situations. Thus, temporal instants come ordered by "before," possible worlds by "accessibility," "transition," or "similarity." We shall stress the intimate

1

relation between types of inference in intensional logic and the kind of semantic pattern presupposed by these. In a sense, this is the most "intrinsic" question generated by our paradigm, concerning the nature of our semantic pictures, and hence of the conceptual enlightenment provided by this kind of semantics. It is precisely because these correspondences are often surprising that intensional semantics has been more than a systematic spelling out of the obvious.

The plan of these notes is as follows. First, some central examples of the more traditional phase of the subject are presented: the logic of tense, modality, conditionals, and their combinations. Then, what is perhaps the most conspicuous recent development is investigated: the shift from "total" to "partial" views of semantic entities and linguistic interpretation. Finally, a connection is made with another recent development in linguistic semantics: intensional notions will be studied as kinds of generalized quantifiers.

Obviously, a selection had to be made in all of this—and I have followed my predilections (and abilities). There is more of an emphasis on propositional than on predicate logic, there is a constant playing down of pretended exclusive rights to "the" philosophical interpretation of the formalism, and, finally, no comprehensive survey of the literature is attempted.

One cannot live in Stanford for long without realizing that "possible worlds semantics" is losing favor with the philosophical community. Evidently, fighting trends is a losing battle, in philosophy as much as in science or fashion. Nevertheless, this course is meant to give an overview of the above *research program*, as distinct from some of the philosophical *ideologies* commonly attached to it. The fund of questions and approaches gathered under the above umbrella, constantly renewing itself since the sixties, is something even the most advanced semanticists should come to terms with; I would hope, terms of endearment.

Concerning the Second Edition (Summer 1987)

One of the surprising things about Intensional Logic is its ability to inspire new applications, even after its reputed demise. Especially in Computer Science, there is a flourishing these days of notions and techniques from possible worlds semantics. A new chapter has been added to describe some of the main developments here. This computational connection also has beneficial effects on Intensional Logic itself: many traditional topics, for instance in the logic of knowledge, have acquired a new impetus.

In addition, the main text has been corrected and expanded, for instance in the discussion of temporality and partiality. At the end, a new chapter has been added on intensional types, reflecting current interest in flexible type theories for logical semantics.

I Classical Theories

Many different types of intensional phenomena have become the subject of special branches of Intensional Logic. Perhaps the most central example is modal logic, but there is also epistemic logic (knowledge and belief), deontic logic (obligation and permission), tense logic, as well as their offspring. For instance, modal logic has served as a model for such diverse theories as the provability logic of arithmetic and the "dynamic logic" of computation (and action in general). Moreover, various combined systems have been investigated. The strategy in these notes has been, not to find a greatest common denominator for all of this, but rather to present the three specific examples which seem to have been richest as research programs: tense, modality and conditionals. We begin with the topic which is generally felt to be the most concrete of the three.

1   Tense and Time

Literature

Prior, A. 1967. *Past, Present and Future*. Oxford: Clarendon Press.
Van Benthem, J. 1983. *The Logic of Time*. Dordrecht: Reidel.

Motivation

The subject of "tense logic" has had a dual motivation. From a philosophical and logical point of view, there is an interest to the calculus of temporal reasoning, breaking with the traditional view that reasoning can only involve timeless eternal propositions. Of the many temporal indicators in natural language, tenses turn out to be reasonably universal and stable. From a linguistic point of view, a logical description of such a ubiquitous and important phenomenon as tense will obviously also be quite welcome.

Some simple examples of temporal reasoning will serve as a point of departure. There is a synonymy between *Dahlia will sob or snore* and *Dahlia will sob or Dahlia will snore*. This ought to be explained by our semantics. Opinions differ as to the synonymy between *Dahlia had lied* and *Dahlia lied*. Our semantics ought to explain the temporal option involved here. And finally, patent non-inferences demand explanation as well, for instance the one from *Dahlia will laugh* and *Dahlia will cry* to *Dahlia will laugh-and-cry*.

Our first step (by no means a negligible discovery) is the introduction of a suitable notation:

$$F\varphi: \quad \varphi \text{ will be the case (at least once)},$$
$$P\varphi: \quad \varphi \text{ was the case (at least once)}.$$

The above examples then become:

$$F(q \vee r) \leftrightarrow Fq \vee Fr, \quad PPq \leftrightarrow Pq?, \quad Fq \wedge Fr \not\rightarrow F(q \wedge r).$$

Notice that the exact derivation of these *logical forms* from the earlier actual *sentences* may already involve some manipulation. Ever since Montague's pioneering work, logicians have become aware that the establishment of such links, traditionally thought of as an art rather than a science, may actually have a logic of its own, that ought to be an explicit part of the semantic enterprise. This point of view seems valid, but we shall continue to sin in these notes.

There are more profound examples of temporal reasoning in the philosophical tradition, where actually something is at stake, such as McTaggart's famous proof of the "Unreality of Time." Many of these have a *modal* flavor as well, but an example seems pertinent, even at this stage.

In the *Master Argument* of Diodorus Cronos (as reconstructed by M. White), Diodorus is reported to have proved the metaphysical principle of "Plenitude" (all possibilities are actual), in the following form:

$$\Diamond \varphi \to \varphi \lor F\varphi.$$

That is, whatever is possible ($\Diamond$) is or will be the case! In a historical analysis, White finds the following premises involved in the argument:

a. $FF\varphi \to F\varphi$,

b. $F\varphi \to G(F\varphi \lor \varphi \lor P\varphi)$ (with "G" standing for "it is always going to be the case": $G\varphi \leftrightarrow \neg F\neg\varphi$),

c. P *true*,

as well as a principle expressing Necessity ($\Box$) of the past:

d. $Pq \to \Box Pq$.

(The Necessity of the Past and Modal-Tense Logic Incompleteness, *Notre Dame Journal of Formal Logic*, 25:1, 1984, pp. 59–71.)

We shall evaluate this argument in the semantics of Chapter 4—with the outcome that Diodorus' argument is valid for a certain metaphysical treatment of possibility and necessity, but not for plausible temporalized accounts of these notions.

Linguistically inclined students were struck by some pleasant analogies between the above temporal sentence operators and tenses in natural language, as tabulated below:

Dahlia lies	q	*Dahlia had lied*	PPq
Dahlia will lie	Fq	*Dahlia will have lied*	FPq
Dahlia lied	Pq	*Dahlia would lie*	PFq

correspondences have been suggested, such as PFP—*would have*, FF—*will be going to*. But eventually, the infinite number of formal combinations is going to outrun the actual (finite) number of tenses: there is "too much." There is also "too little": tenses such as the present perfect and progressive remain unaccounted for. But then, the formalism can be extended in due course to include further operators.

The Basic Formal Semantics

The above notation may be thought of as a formal language, in the simplest case, a propositional language of the usual kind, with added operators F, P, G (as above), and also H ("always in the past," $H = \neg P \neg$).

Models $M = \langle T, <, V \rangle$ consist of a "flow of time" $\langle T, < \rangle$, (i.e., a set of "moments" ordered by "earlier than" or "before") with a "valuation" V giving, for each proposition letter q, the set $V(q)$ of times when q is true. Alternatively, V provides a family of ordinary propositional valuations ("snapshots"), one at each moment in time. One can think here of various pictures of time: linear sequences, or perhaps branching trees, with changing events.

As usual, a recursive truth definition fixes the interpretation of the symbolism from the proposition letters upward:

$$M \models \varphi[t] \qquad (\text{"}\varphi \text{ is true in } M \text{ at } t\text{"}).$$

The clauses are as follows:

$$M \models p[t] \quad \text{iff} \quad t \in V(p),$$
$$M \models \neg\varphi[t] \quad \text{iff} \quad not \; M \models \varphi[t],$$
(and likewise for $\wedge/and$, $\vee/or$, $\rightarrow/if\ then$, etc.),
$$M \models F\varphi[t] \quad \text{iff} \quad for\ some\ t' > t,\ M \models \varphi[t']$$
$$M \models P\varphi[t] \quad \text{iff} \quad for\ some\ t' < t,\ M \models \varphi[t']$$
(and likewise for G and H).

If one wants to speak here of "the denotation" of a sentence φ in a model M, the obvious candidate is not one truth value, but a family of these: one for each world (Montague-style). Equivalently, one could say that the denotation of φ is the set of all points in time where φ is true (the common practice in technical tense logic). Notice that this view does not automatically commit one to the converse: that every set of points in time is an admissible "proposition." Indeed, models $\langle T, < \rangle$ with restricted ranges of denotations for sentences have been used extensively in technical research.

Digression. it may be instructive to realize that the view of sentences as denoting a *truth value* occurs in only one tiny corner of classical logic, viz. "local" propositional semantics. But even in propositional logic already, a sentence φ is often taken to denote all valuations making it true (if you wish, all "situations" where it holds). Moving to predicate logic, a proper Fregean compositional setup makes it imperative to let formulas denote sets of variable assignments. (Properly viewed, this is also what we have done in the above tense-logical case.) When special care is needed with complexity (as in set-theoretic meta-arguments), these will even be partial rather than total assignments. In other words, many monolithic philosophical views of "denotation" have been far removed from the realities (and flexibility) of logical practice.

When the temporal operators are added to a predicate logic, the above picture will become more detailed, locating, at each point $t \in T$, a whole Tarskian structure consisting of a domain of discourse and interpretations for basic predicates. In particular, this picture allows variation in domains over time, as well as changes in the behavior of individual objects. Such variation is needed to account for the various readings of, say,

$$\textit{Every man will do his duty}: \quad \forall x(man(x) \rightarrow F\,duty(x)),$$
$$F\,\forall x(man(x) \rightarrow duty(x));$$

$$\textit{Every boy will have heard Dahlia}: \quad \forall x(boy(x) \rightarrow FP hear(x,d)),$$
$$F\,\forall x(boy(x) \rightarrow P hear(x,d)),$$
$$FP\,\forall x(boy(x) \rightarrow hear(x,d)).$$

A more radical way of "temporalizing" our ontology would be to dismiss the above stable objects, and take individuals to be "lifelines" assigning to each moment the corresponding manifestation of the individuals. (Compare Montague's "individual concepts.") This view, though intriguing, has never become predominant.

Evidently, the semantic scheme presented here can be used to interpret many other temporal expressions. The above is just the simplest system in existence, be it a fairly typical one.

There is a minimal logic of the above semantic scheme, consisting of all those formulas which are valid in all models at all points in time. One example is the earlier equivalence $F(q \vee r) \leftrightarrow Fq \vee Fr$. (Basically, the existential quantifier in the semantic clause for F distributes over disjunction.) On the other hand, for example, $Fq \wedge Fr \rightarrow F(q \wedge r)$ has obvious counterexamples. The *minimal tense logic* thus obtained can be axiomatized as follows:

a. all propositional tautologies;

b. the definitions:

$$F\varphi \leftrightarrow \neg G \neg \varphi,$$
$$P\varphi \leftrightarrow \neg H \neg \varphi;$$

c. the tense-logical axioms:

$$G(\varphi \rightarrow \psi) \rightarrow (G\varphi \rightarrow G\psi),$$
$$H(\varphi \rightarrow \psi) \rightarrow (H\varphi \rightarrow H\psi),$$
$$\varphi \rightarrow GP\varphi, \quad \varphi \rightarrow HF\varphi;$$

d. the rules of inference:

$$\varphi, \varphi \rightarrow \psi / \psi \qquad \text{(Modus Ponens)},$$
$$\varphi / G\varphi, \ \varphi / H\varphi \qquad \text{("Eternity")}.$$

But, further principles become valid as soon as one imposes reasonable constraints on the temporal order. Indeed, precise *correspondences* arise between tense-logical axioms and ordering conditions. To state these, the following notion of truth is needed, depending only on the ordering pattern, ignoring accidental features of any particular valuation:

- $\mathbf{T}\ (=\langle T, < \rangle) \models \varphi[t]$ if $\langle T, <, V \rangle \models \varphi[t]$ for all valuations V,

- $\mathbf{T} \models \varphi$ if $\mathbf{T} \models \varphi[t]$ for all $t \in T$.

An example is provided by the earlier putative principle $PPq \leftrightarrow Pq$. We have:

$$\mathbf{T} \models PPq \rightarrow Pq\,[t] \quad \text{iff} \quad \forall x < t\ \forall y < x\colon\ y < t \quad (transitivity),$$
$$\mathbf{T} \models Pq \rightarrow PPq\,[t] \quad \text{iff} \quad \forall x < t\ \exists y < t\colon\ x < y \quad (density).$$

Thus, we see which consequences endorsing the above equivalence of simple past and past perfect would have for our picture of time.

Other examples occur in the mentioned version of Diodorus' Master Argument:

- $FFq \rightarrow Fq$ defines *transitivity* as well,

- $Fq \rightarrow G(Fq \vee q \vee Pq)$ defines *right-linearity*
  (i.e., $\forall t\, \forall x > t\ \forall y > t\ (y < x \vee y = x \vee x < y)$), while

- $P\ true$ defines *left-succession* ($\forall t\, \exists x\colon x < t$).

Conversely, interesting tense-logical axioms have also been discovered in attempts to match existing conditions on the temporal order. For instance, in order to describe *discreteness*

$$\forall t\colon\ \exists x > t\ \forall y < x\ (y = t \vee y < t),$$
$$\forall t\colon\ \exists x < t\ \forall y > x\ (y = t \vee y > t),$$

"Hamblin's Axiom" was invented:

$$(q \wedge Hq) \rightarrow FHq, \quad (q \wedge Gq) \rightarrow PGq.$$

For a survey of this area, see J. van Benthem, Correspondence Theory, in D. Gabbay and F. Guenthner, eds., *Handbook of Philosophical Logic*, Vol. II, Reidel, Dordrecht, 1984.

Thus, through a study of actually proposed "valid" patterns of inference, one can form a conception of their presupposed picture of Time. This is the direction of modeling various temporal logics. Conversely, one may start with some preferred picture of Time, say, that of the *integer* or *real* line, and ask for all valid inferences there. For most well-known structures, such axiomatizations had been found by 1970. Actually, there is no general logical reason why all these structures should have axiomatizable theories at all. And, at least for temporal *predicate* logic, there is a well-known warning example (discovered independently by

Lindström and Scott): the tense predicate logic of the real time line is not effectively axiomatizable.

Of course, questions of correspondence and axiomatization are not the only queries arising in this semantics. For instance, an obvious question is, on any given temporal structure, what happens to the potentially infinite number of "tenses," that is, sequences of operators F, P, G, H in our language. A good exercise to become familiar with the peculiarities of our formalism is to prove Hamblin's *Fourteen Tenses Theorem*:

> On the real time axis, there are only 14 logically distinct sequences of operators.

(Hint: use simple collapsing principles such as "$PP = P$," "$GG = G$," as well as more exotic ones, such as "$FHF = F$," "$FHP = HP$." Moreover, cut down on calculations by exploiting symmetries.) Thus, the issue whether our formalism can match natural language tenses becomes more interesting, with possibly different answers in different semantic settings.

Finally, there is a logical interest to the study of flows of time $\langle T, < \rangle$, independently from any particular language being interpreted. For instance, many intuitions that we have concerning Time do not correspond to simple inferences, but are of a more global nature. One famous example is *Homogeneity*: "all points in Time are alike." It has the following technical formulation: "every point in T can be mapped onto any other one by some order-preserving automorphism of $\langle T, < \rangle$." Such intuitions impose more global constraints on our class of semantic models, with less "direct" effects than, say, the earlier transitivity or asymmetry. For instance, Homogeneity tells us (among other things) that the temporal order cannot change its behavior: for example, it is either dense or discrete throughout. Actually, the statement of such general intuitions, beyond what is needed for immediate "engineering" purposes, is a discernible recent trend in various areas of semantics.

Further Developments

The above paradigm has been extended repeatedly to account for further temporal phenomena. One notable direction here has been the introduction of auxiliary reference points. For instance, the permanence of "now," no matter how deeply embedded in a sentence, requires the continuing availability of the original moment of utterance, even while operators P and F are being unpacked, shifting the current point of evaluation. Thus, Kamp introduced "double indexing": φ is true in M at t, t_0; with a resetting function for "now." NOW φ is true in M at t, t_0 if and only if φ is true in M at t_0, t_0. Further moves in this direction have been proposed by Vlach, Åqvist and Guenthner, and by Gabbay—until a system arose whose temporal operators looked remarkably like the *Quine operators* for a variable-free predicate logic. By this time, the tense-logical formalism had become as strong as a full-blown two-sorted predicate logic, consisting of an ordinary predicate logic with an added domain of temporal items, freely allowing for quantification over the latter. (For instance, Fq corresponds to $\exists t > t_0 \ Qt$,

$$\underline{F \, \forall x(\text{NOW } girl(x) \rightarrow woman(x))}$$

to

$$\exists t > t_0 \ \forall x \ (girl(t_0, x) \rightarrow woman(t, x)),$$

while, for example, $\exists t < t_0 \ \exists t' > t_0 \ \forall s(t < s < t' \rightarrow rains(s))$ expresses the progressive *it is raining*.) A whole philosophical and methodological debate has raged about the relative virtues of these two formalisms.

The above two-sorted predicate logic with temporal parameters can be regarded as a kind of "descriptive limit" for temporal constructions. An early result is "Kamp's Theorem" showing that (modulo some assumptions on the temporal order) actually two constructions suffice for obtaining its full power:

$$\text{SINCE } \varphi\psi: \quad \exists t' < t_0(\varphi_{t'} \wedge \forall t'' \in (t', t_0)\psi_{t''}),$$
$$\text{UNTIL } \varphi\psi: \quad \exists t' > t_0(\varphi_{t'} \wedge \forall t'' \in (t_0, t')\psi_{t''}).$$

Thus, in a sense, natural language attains maximal power of temporal expressibility using these two notions.

Finally, the perspective presented here has some obvious descriptive limitations. (The point, of course, is not that its students failed to notice these, but rather that they hoped to gain insights from a judicious idealization, as so often in science.) For instance, there are the "missing tenses," such as present perfect or progressive. Now, the latter can be given a new operator, but the former presents a more subtle problem. Its truth conditions seem to be identical to those for the simple past (P), and yet there is a difference. Reichenbach treated this distinction, already in the forties, using a so-called "reference point," introducing a perspective upon the event described: the simple past (*Dahlia sang*) has its perspective located at that past event, the present perfect (*Dahlia has sung*) may look at that same event from the present. Attempts to incorporate this idea into the earlier framework have not been quite successful. (We shall find a more recent attempt in the interval tense logic of Chap. 6.)

Another famous problem is the deictic use of tenses, as exemplified in Partee's *I did not turn off the stove*, which refers to some specific past time. Here, neither reading in the above formalism seems appropriate: $P\neg \ turn \ off$ is too weak and $\neg P \ turn \ off$ much too strong. (But perhaps, this just points at a "specific" reading of the existential quantifier in the $P\neg$-version—a phenomenon that has to be taken seriously even in ordinary predicate logic.)

Finally, the merest look at actual language patterns shows that the above "tenses" are actually compound uses of bona fide *tense* (in English morphology: only *present* and *past*) and *temporal auxiliaries* (*have, will, be*). A more sensitive description might produce "unanchored" correlates first for, say, *walk, have walked,* *[*will*] *walk, be walking*: using the tense to anchor these in real time.

2 Modality

Literature

Hughes, G., and M. Cresswell. 1968. *An Introduction to Modal Logic.* London: Methuen.

Motivation

The logical study of the philosophical modalities ("necessarily," "possibly") dates back to Aristotle. The immediate cause for the birth of modern "modal logic" was more special, springing from dissatisfaction with the Frege-Russell truth table treatment of *implication*. Where propositional logic makes $\varphi \rightarrow \psi$ equivalent to $\neg(\varphi \wedge \neg\psi)$ (φ does not occur with *not-ψ*), C. Lewis thought it had at least the modal force $\neg\Diamond(\varphi \wedge \neg\psi)$ (φ *cannot* occur with *not-ψ*), or equivalently $\Box(\varphi \rightarrow \psi)$, ($\varphi$ *necessarily* implies ψ materially). Gradually, the study of $\Diamond$ and $\Box$ by themselves became dominant, with studies of "implication" becoming a separate subject (see Chap. 3).

There has been less direct linguistic motivation in this area. *Must* and *may* are to some extent natural language expressions of necessity and possibility, but there are also some divergences. For instance *Cynthia must be sleepwalking* may be an inference from hearing footsteps on the stairs, with less force than *Cynthia is sleepwalking*, whereas the principle $\Box\varphi \rightarrow \varphi$ is axiomatic in modal logic. On the other hand, this example also shows an analogy: necessity says something about our *reasons* for holding a proposition. Indeed, Frege held (with Kant) that qualifications of modality add nothing to the content of a judgment, telling us only something about our attitude towards it. Thus, it is not obvious that one would need a logical semantics for modality at all. Instead of pursuing this issue in an a priori manner, we shall present a semantics and comment upon its uses. Even with a less than immaculate philosophical conception, this *kind* of logical setup has become very influential as a prototype for the semantics of intensional phenomena in natural language.

The philosophical inspiration for developing modal logic has both "local" forms—analysis of particular modal arguments from the tradition—and more "global" ones: trying to (re)construct coherent philosophical views of modality.

Example 1 (Aristotle's "Sea Battle Argument"):

"If I give the order to attack (p), then, necessarily, there will be a sea battle tomorrow (q). If not, then, necessarily, there will not be one. Now, I give the order, or I do not. Hence, either it is necessary that there is a sea battle tomorrow, or it is necessary that none occurs."

This famous logical argument for determination of the future (the admiral does not "really" have a choice) has been revived again and again in the the philosophical tradition, in this century by J. Lukasiewicz and R. Taylor. One first analysis employs a modal notation similar to that in Chapter 1, with the following *two* versions arising:

$$
\begin{array}{ll}
p \to \Box q & \Box(p \to q) \\
\neg p \to \Box \neg q & \Box(\neg p \to \neg q) \\
p \vee \neg p & p \vee \neg p \\
\hline
\Box q \vee \Box \neg q & \Box q \vee \Box \neg q
\end{array}
$$

The crucial point is the scope of the "necessarily" in the first premises. In the first reading (narrow scope), the argument is valid, but the (strong) premises beg the conclusion. In the second reading (wide scope), the premises are plausible, but the conclusion does not follow. (These assertions are to be made good in our eventual semantics, of course.)

Example 2 (Quine's "Mathematical Cyclist"):

"Mathematicians are necessarily rational, but not necessarily bipeds. Cyclists are necessarily bipeds, but not necessarily rational. But then, what about the mathematician cyclist Paul K. Zwier: he is both necessarily rational and not necessarily rational ... and his motion is as contradictory as his mind?" This is one of Quine's arguments against the very consistency of a modal logic applying to predicates and individuals. Again, to defuse the attack, some distinctions are to be made. For instance, the very first assertion is three-ways ambiguous.

$$\forall x(mathematician(x) \to \Box rational(x))$$

ascribes necessary rationality to actual mathematicians (a so-called "de re" assertion of modality). But, there is also the "de dicto" assertion

$$\Box \forall x(mathematician(x) \to rational(x))$$

about the proposition that mathematicians are rational; and finally, there is even an intermediate possibility

$$\forall x \,\Box(mathematician(x) \rightarrow rational(x)).$$

An aside: the exact justification for postulating such different readings on the basis of the above simple syntactic form remains to be spelled out. Nevertheless, the de dicto reading seems obviously present, even though the actual syntactic pattern seems to favor the de re reading. (Compare *Mathematicians are usually ill-tempered*, which seems to have only a marginal de re reading.)

The development of entire modal ontologies hand in hand with a semantics for the above formalism has given rise to an extensive literature. Some influential publications are A. Plantinga, *The Nature of Necessity*, Clarendon, Oxford, 1974; A. Prior and K. Fine, *Worlds, Times and Selves*, The University of Massachusetts Press, Amherst, 1977; and S. Kripke, *Naming and Necessity*, Harvard University Press, Cambridge, Massachusetts, 1980.

Possible Worlds Semantics

As before, our basic language is that of propositional or predicate logic, enriched with the modal operators $\Box$ and $\Diamond$. The semantic structures are going to be formally identical to those in tense logic, but with a more esoteric "conceptual tale." In the propositional case,

$$M = \langle W, R, V \rangle,$$

with W thought of as a set of *"possible worlds,"* R a relation of *"accessibility,"* and V a *valuation* as before. A simple example is the "chessboard": worlds are all possible configurations of pieces on the board; in any of these, the accessible worlds are those that can still be obtained by further play within the rules of the game. Thus, R encodes constraints on our modal options, much as $<$ did for Time. Accordingly, the crucial clauses in the truth definition become:

$$M \models \Box\varphi\,[w] \quad \text{iff} \quad \textit{for all } v \textit{ with } Rwv, \; M \models \varphi[v],$$
$$M \models \Diamond\varphi\,[w] \quad \text{iff} \quad \textit{for some } v \textit{ with } Rwv, \; M \models \varphi[v].$$

This idea of necessity as truth "in all possible worlds" has a conceptual prehistory dating back to Leibniz (and far beyond). Nevertheless, this semantic picture is more abstract than the earlier temporal one, as it does not evoke spontaneous concrete pictures of a "modal universe."

Some examples of how this semantics works will make it more concrete.

Example 3 (Validity of Modal Distribution):

The principle $\Box(\varphi \rightarrow \psi) \rightarrow (\Box\varphi \rightarrow \Box\psi)$ is universally valid ("entailments of necessary truths are themselves necessary truths"). For, if

$\Box(\varphi \to \psi)$ holds at w, and so does $\Box\varphi$, then, at each accessible v, both $\varphi \to \psi$ and φ hold: implying ψ.

Example 4 ("Scope Differences Explained"):

The following two pictures establish the independence of the two readings $p \to \Box q$, $\Box(p \to q)$ in the Sea Battle argument:

(i)
$$
\begin{array}{cc}
1 & 2 \\
\bullet \longrightarrow \bullet \\
p & \neg p \\
& \neg q
\end{array}
$$
in world 1: $\Box(p \to q)$ is true,
$p \to \Box q$ is false;

(ii)
$$
\begin{array}{cc}
1 & 2 \\
\bullet \longrightarrow \bullet \\
\neg p & p \\
& \neg q
\end{array}
$$
in world 1: $\Box(p \to q)$ is false,
$p \to \Box q$ is true.

Again, statements φ will correspond to sets of possible worlds ("propositions"): $\{ w \in W \mid M \models \varphi[w] \}$. The converse need not hold, as was observed before in Chapter 1. (A comment for the technically minded: With the full range of subsets of W, modal logic has features of a second-order logic. With restrictions, one has Henkin's "general model" scheme, which, in a sense, makes the higher-order logic into a many-sorted first-order one. This possibility of "flattening" modal theories such as, say, Montague's intensional logic, has been pointed out by several observers.)

With the modalities superimposed upon a predicate logic, possible worlds will again acquire an internal structure with individuals and predicates. The truth definition then offers several options, most of which have found defenders in the literature. For *atomic assertions*, a decision is needed as to the interpretation of statements at w involving individuals not occurring in w's universe. (Here, we shall leave these uninterpreted.) Next, the *range of individual quantifiers* is to be determined. Most accounts let them range over the world-restricted universes:

$M \models \forall x\, \varphi[w, A]$ iff   *for all individuals d in the universe of w:*
$$M \models \varphi[w, A^x_d].$$

Here A is an assignment of values in w's universe to free variables, with A^x_d the obvious modification. (An alternative would be to let $\forall$ and $\exists$ range over all individuals, whether "actual" or just "possible" in w.) Finally, statements of necessity may now refer to specific individuals, which may or may not be present in alternative worlds. Out of various policies here, the following one seems reasonable:

$M \models \Box\varphi\,[w, A]$ iff   *for all worlds v with Rwv whose universe*
contains all individuals $A(x)$ (for x free in φ):
$$M \models \varphi[v, A].$$

A useful exercise to explore the finer points of this semantics is to investigate the (possible) validity of the following de re- / de dicto-interchange principles:

(1) $\Box \forall x A x \rightarrow \forall x \Box A x$,

(2) $\forall x \Box A x \rightarrow \Box \forall x A x$ (the so-called "Barcan Formula"),

(3) $\Box \exists x A x \rightarrow \exists x \Box A x$,

(4) $\exists x \Box A x \rightarrow \Box \exists x A x$.

It will be seen that only (1) is valid, with the others making varying demands upon the combined structure of alternatives and universes:

$$(2) \text{ corresponds to } \forall wv(Rwv \rightarrow U_v \subseteq U_w),$$
$$(4) \text{ corresponds to } \forall wv(Rwv \rightarrow U_w \subseteq U_v),$$

and (3) expresses a higher-order condition. (Note that (3) would be valid if individuals were thought of as "individual concepts," such as *the winner*.) Thus again, modal inferences translate into structural conditions on our semantic picture.

Logical Theory

There is a vast technical literature on this semantics, witness B. Chellas, *Modal Logic: An Introduction*, Cambridge University Press, Cambridge, 1980; D. Gabbay, *Investigations in Modal and Tense Logics*, Reidel, Dordrecht, 1976; G. Hughes and M. Cresswell, *A Companion to Modal Logic*, Methuen, London, 1984; and J. van Benthem, *Modal Logic and Classical Logic*, Bibliopolis/ The Humanities Press, Napoli/ Atlantic Heights, 1985.

One striking feature is the multiplicity of modal logics, beyond the *minimal modal logic K* (in honor of Kripke) validated by the above scheme without restrictions on R:

a. all propositional tautologies,

b. the definition $\Diamond \varphi \leftrightarrow \neg \Box \neg \varphi$,

c. $\Box(\varphi \rightarrow \psi) \rightarrow (\Box \varphi \rightarrow \Box \psi)$,

d. Modus Ponens and Necessitation (from φ to $\Box \varphi$).

(A useful exercise in deduction within this system is provided by the claims made in the "Smith Example," at the end of this chapter.)

Beyond K, opinions have diverged as to reasonable further assumptions about the modalities. Usually, these have taken the form of "reduction principles," relating one modality (i.e., sequence of modal qualifications) to another. Well-known examples are $\Box \varphi \rightarrow \varphi$, $\Box \varphi \rightarrow \Box \Box \varphi$ (together, these turn K into the modal logic $S4$), and $\Diamond \Box \varphi \rightarrow \varphi$ (with $S4$, this becomes $S5$). But also, more involved examples have turned up in various contexts, such as the Geach Axiom $\Diamond \Box \varphi \rightarrow \Box \Diamond \varphi$ ($S4.2$), or the McKinsey Axiom $\Box \Diamond \varphi \rightarrow \Diamond \Box \varphi$ ($S4.1$).

One of the most striking virtues of possible worlds semantics was its analysis of these syntactic logics in terms of simple, well-understood conditions on their models. For example, $\Box\varphi \rightarrow \varphi$ wants the alternative relation R to be *reflexive*, $\Box\varphi \rightarrow \Box\Box\varphi$ *transitive*, and $\Diamond\Box\varphi \rightarrow \varphi$ *symmetric*. (Thus, *S5* demands an equivalence relation on W.) The Geach Axiom makes R *directed*:

$$\forall w\, \forall xy((Rwx \wedge Rwy) \rightarrow \exists z(Rxz \wedge Ryz)).$$

The McKinsey Axiom is more complex, however, and fails to correspond to a simple first-order condition on alternatives. (A useful special case will be needed in Chap. 8.) On *transitive* models, the McKinsey Axiom corresponds to the existence of end-points:

$$\forall w\, \exists x(Rwx \wedge \forall y(Rxy \rightarrow y = x)).$$

Thus, one could now chart the connections between the various proposed logics of modality in terms of structural connections between their models. This is one of the pleasant cases where a semantics provides insights beyond faithful recording of known truth-conditional intentions.

Formal and Applied Semantics

The above formal theory has been made the basis of a philosophical ontology, or "world view," by S. Kripke, D. Lewis, A. Plantinga, K. Fine, J. Hintikka, and others. It is this *applied* paradigm which is often referred to as "possible worlds semantics." In this case, one has to form a conception of what possible worlds *are*: alternative physical universes, storybooks, etc. Moreover, the alternative relation will need some further conceptual interpretation. As is to be expected in philosophy, one finds a whole spectrum here, from "realists" about possible worlds to "instrumentalists."

Halfway between a full-fledged philosophical ontology and the bare formalism are accounts reading necessity as derivability—from some set of principles (logic, physics, morality, etc.). In each case, the "possible worlds" will be the models for those principles, and one gets a hierarchy:

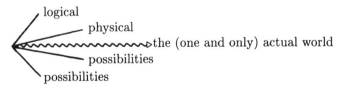

Notice the inversion: stronger background theories, fewer possible worlds.

One particular problem faced by these explications is Quine's Challenge: What does it mean to ascribe some property necessarily to an individual (de re), and are there any non-trivial instances?

But, the possible worlds formalism does not stand or fall with these philosophical world views. It has also received more mundane practical interpretations and applications, of independent interest.

Example 5 (Provability Logic):

If one interprets $\Box\varphi$ as "φ is provable in Peano Arithmetic," valid principles from the metamathematics of arithmetic turn into familiar modal axioms, notably $\Box(\varphi \to \psi) \to (\Box\varphi \to \Box\psi)$, $\Box\varphi \to \Box\Box\varphi$, "Necessitation," and, most striking, Löb's Axiom:

$$\Box(\Box\varphi \to \varphi) \to \Box\varphi.$$

(The fact is that $\Box\varphi \to \varphi$ is not generally provable in PA—even though this modal axiom is intuitively true. For the full story, see G. Boolos, *The Unprovability of Consistency*, Cambridge University Press, Cambridge, 1979. Notice that $\Box\neg true \to \neg true$ would be equivalent to the consistency statement $\neg\Box false$.) Löb's Axiom had already been studied in modal logic, and its correspondence behavior was known:

$$\langle W, R \rangle \models \Box(\Box p \to p) \to \Box p \quad \text{iff} \quad R \text{ is } transitive \text{ and}$$
$$upwards \ well\text{-}founded.$$

(Proving this equivalence will make the reader thoroughly familiar with the less routine aspects of modal correspondences.)

Thus, we have a semantics for the Peano provability logic, for which Solovay found a complete description combining techniques from modal logic and recursion theory. (It is only fair to add, however, that the precise arithmetical *meaning* of possible worlds models in this metamathematical setting is not yet fully understood.)

Example 6 (Dynamic Logic):

Interpreting $\Box_\alpha\varphi$, for some program α, as "φ holds after every execution of α" makes the modal language a medium for describing behavior of computer programs. Worlds will now be states of a computer, and accessibility relations (indexed by programs) give possible transitions. An initial result of some note is the axiomatization of the minimal logic in this area, when α belongs to the "regular programs" generated using sequencing (;), indeterminate choice ($\cup$), and finite iteration (Kleene *). (This result was not an obvious corollary of earlier modal completeness theorems, since the Kleene star has an infinitary flavor.)

A good initial survey is found in V. Pratt, Applications of Modal Logic to Programming, *Studia Logica* 39, 1980.

Finally, it remains to be repeated that various other intensional logics have been developed on the above pattern, such as Hintikka's treatment of knowledge and belief ("epistemic logic"):

$$K_a\varphi(\text{"}a \text{ knows that } \varphi\text{"}) \quad \text{if} \quad \varphi \text{ holds in all of } a\text{'s epistemic}$$
$$\text{alternatives.}$$

A very thorough reference on the latter topic is W. Lenzen, *Glauben, Wissen und Wahrscheinlichkeit*, Springer, Vienna, 1980. Another notable example is Von Wright's "deontic logic" of obligation and permission.

Appendix: A Forgotten Paradigm

Possible worlds semantics captured one dominant view of modality. But still in the thirties, C. Lewis had competitors, stressing different conceptual priorities. For instance, H. B. Smith searched for a modal logic that would validate two broad intuitions concerning modality:

no two modalities are to be provably equivalent   (Distinction),
provable implication arranges all modalities linearly   (Hierarchy).

(Recall that "modalities" are sequences of modal operators.) For a reference, compare the exposition in P. Rosenbloom, *The Elements of Mathematical Logic*, Dover, New York, 1950, pp. 61–63.

A "virtual impossibility" result is the following:

Theorem: On the basis of the minimal modal logic K, Smith's conception cannot be realized.

Proof: We have either $\varphi \to \Box\varphi$ or $\Box\varphi \to \varphi$ (Hierarchy). In the former case, $\Box\varphi \leftrightarrow \Box\Box\varphi$ becomes derivable in K, contradicting Distinction. In the latter case, one has either $\varphi \to \Diamond\Box\varphi$ (and the forbidden $\Diamond\varphi \leftrightarrow \Diamond\Box\Diamond\varphi$ becomes derivable), or $\Diamond\Box\varphi \to \varphi$ (and one ends up with $\Box\Diamond\varphi \leftrightarrow \Box\Diamond\Box\Diamond\varphi$). All roads are blocked. ∎

Thus, any attempt to semanticize Smith's ideas is bound to challenge even the minimal modal logic (for which precedents exist, even within the possible worlds paradigm).

The above result illustrates the dangers of logical vetoes. During the course, Tore Langholm found a modelling for Smith's conception after all. (See his paper H. B. Smith on Modality: a Logical Reconstruction, *Journal of Philosophical Logic*, to appear.) Dropping the K-rule of Necessitation, one can have a logic given by all principles

$$\Box\varphi \to \varphi, \quad \Diamond\Box\varphi \to \varphi, \quad \Diamond\Diamond\Box\varphi \to \varphi, \quad \text{etc.,}$$

which has the following properties:

- it is complete with respect to models $\langle W, R, w_0 \rangle$ with a distinguished actual world $w_0 \in W$, satisfying the condition:
  if $w_0 R x_1 R \ldots R x_n$, then $R x_n w_0$ $(n = 0, 1, 2, \ldots)$,

- it satisfies both Distinction and Hierarchy,

- upon the addition of Necessitation, it collapses into $S5$.

One interesting feature is the order of modalities in this semantics: they form an *unbounded linear dense* pattern (like the rational numbers). This gradual shift between modalities, as opposed to discrete steps, is another feature of various traditional views of modality.

3 Conditionals

Literature

Sosa, E., ed. 1975. *Causation and Conditionals.* Oxford: Oxford University Press.

Harper, W., et al., eds. 1981. *Ifs.* Dordrecht: Reidel.

Motivation

Dissatisfaction with material implication led to the modal entailment tradition of explicating conditional statements *"if φ, then ψ."* This direction may be described as being "vertical," diverging from classical $\rightarrow$ when iterations are concerned. (Notably, a "fallacy of relevance," such as the principle $\varphi \rightarrow (\psi \rightarrow \varphi)$, would no longer be valid.) But, as long as attention is restricted to single implications $\varphi \Rightarrow \psi$ ("$\Rightarrow$" now being our general implication symbol), with perhaps Boolean operators inside φ, ψ, all modal logics from K upward have the same theory for $\varphi \rightarrow \psi$ and $\Box(\varphi \rightarrow \psi)$.

The tradition to be presented now has a "horizontal" quarrel with material implication, challenging its property of *monotonicity*:

$$\varphi \Rightarrow \psi \quad \text{only if} \quad \varphi \wedge \chi \Rightarrow \psi \qquad \text{(Strengthening of the Antecedent).}$$

One counterexample is the following:

> φ: I put sugar into my coffee, ψ: it tastes fine,
> χ: I put diesel oil into my coffee.

The intuitive explanation seems to be that we assented to $\varphi \Rightarrow \psi$ *ceteris paribus* (or "under normal circumstances"), while χ introduced an unenvisaged marginal eventuality. This raises the issue of our assenting to rules of non-universal validity—a pervasive phenomenon in ethics (the term "ceteris paribus" is a juridical one) and even in science.

23

Philosophers of science encountered this kind of conditional in their study of various types of scientific discourse. Carnap found it underlying *dispositional* statements. For example, "soluble" means "dissolves, if immersed in water," with a ceteris paribus "*if*." Goodman found nonmonotone conditionals when analyzing the problem of *counterfactual* statements. "*If the match had been lighted, there would have been an explosion*," but certain unusual circumstances could nullify this regularity. One reason for the philosophical interest in this notion has been its role in the cluster of concepts: *lawlike generalization, generalization supporting counterfactuals, physically necessary statement, causal connection.* For instance, not every true universal statement supports counterfactuals, but lawlike ones do. Unfortunately, also conversely, an explication of counterfactuals seems to presuppose some notion of bona fide general law. But then, as usual in philosophy, one can at least chart the problems in their relationships.

There is also a growing linguistic interest in the semantics of the various kinds of conditional found in ordinary language. There are some broad technical oppositions here, between "individual" and "generic" conditionals, or "indicative" and "subjunctive" ones—but we shall aim at an underlying minimal framework which is still neutral with respect to these specializations. (This strategy is explained rather nicely in A. Kratzer's paper The Notional Category of Modality, in H.-J. Eikmeyer et al., eds., *Words, Worlds and Contexts*, W. de Gruyter, Berlin, 1981.)

The Ramsey Test

As so often in exact philosophy, the heart of the matter is already found in a short paper by F. P. Ramsey in the thirties. According to Ramsey, the following test determines if a conditional $A \Rightarrow B$ is true on the basis of a stock of beliefs T:

Add the antecedent A to T. In case the result becomes inconsistent, make the *minimal revision* needed to accommodate A consistently. Then, check if B follows from the latter modified stock.

Until the sixties, syntactic explications of this view were prevalent (or rather: versions without an *explicit* semantics)—after the advent of possible worlds semantics in modal logic, explicit semantic accounts started appearing. Here is a "syntactical," or proof-theoretic version of the Ramsey Test:

A set of sentences T validates a conditional $A \Rightarrow B$ if, for *every* set X which is *maximal* with respect to the property "$T \vdash X$ and $X + A$ is consistent," $X + A \vdash B$.

Here, $\vdash$ stands for derivability in standard logic. Notice that *all* minimal revisions are considered, as there need not be a unique one.

Now, if A is consistent with T, the above reduces to $T \vdash A \to B$, which is reasonable. But, if A is inconsistent with T, triviality threatens (at least, in a propositional setting with finitely many atomic statements): $A \to B$ must be universally valid! (For otherwise, let Y be any maximally consistent proposition implying $A \wedge \neg B$. Then the disjunction of Y with the conjunction of T forms a set X as described above, without $X + A$ (i.e., Y) $\vdash B$. ∎) Thus, the above idea must be handled with care. More sophisticated explications of the Ramsey Test, employing a preference ranking on statements in T (as a measure of "minimal" revision) may be found in P. Gärdenfors, An Epistemic Approach to Conditionals, *American Philosophical Quarterly* 68, 1981, pp. 203–211.

Digression: Non-classical Notions of Deducibility. One way of reformulating the above scheme is by referring to *subsets* X of T rather than subsets of its deductive closure. Interestingly, validity then becomes dependent on the *presentation* of T. For instance, $T = \{p\}$ does not validate $\neg p \Rightarrow q$ (its only eligible X being the empty subset), but the deductively equivalent $T' = \{p, p \vee q\}$ does (as $p \vee q, \neg p \vdash q$). This is just one of many interesting features that arise once we consider various non-standard notions of "consequence" from a set of premises; studied quite frequently in the philosophy of science (Carnap, Hempel), but regrettably scarce in mainstream logic. Another illustration is the *nonmonotonicity* of the above notion. For example, $\{p \vee q\}$ validates $\neg p \Rightarrow q$, but its consistent extension $\{p \vee q, \neg q\}$ no longer does. (Actually, the *ternary* relation "A implies B in context T" is quite congenial to recent work on this phenomenon in Artificial Intelligence.) Finally, on the interpretation given here, one may also view the problem as one of inference from a possible *inconsistent* theory $T + A$—the proposal being to call B derivable if it follows from every maximally consistent subset of $T + A$. Again, this happens to be precisely one of the modern proposals in the area of "dialectical logic." Thus, the contemporary semantic approach can still profit from a study of some of the subtleties expressed in an earlier paradigm.

The End of the Syntactical Phase

Perhaps the nicest exposition of the syntactic perspective is found in Rescher's paper in the Sosa volume. Our beliefs (theories, prejudices) T come in a *hierarchy* of statements to be given up, say, in the order of laws/auxiliary hypotheses/facts. The test then becomes: to add A to T, removing statements from T to obtain consistency, starting where it hurts least. Example: *If the match had been lighted (L), there would have been an explosion (E).* Let us assume a background theory with one law: $L \wedge O \rightarrow E$, where O is the auxiliary hypothesis that *oxygen was present*, together with the facts $\neg L, \neg E$. It may be checked that $L \Rightarrow E$ is indeed validated.

Difficult cases arise with "symmetries," such as in Quine's example: *If Bizet and Verdi had been compatriots, Bizet would have been Italian / Verdi would have been French?* Rescher would invoke pragmatical considerations here, to break the tie.

A Semantical Explication

Possible worlds explications for the above ideas were proposed by R. Stalnaker and D. Lewis. As the latter version seems more flexible, it will be preferred here.

Consider a propositional language with an added operator $\Rightarrow$. Models will now be structures $M = \langle W, C, V \rangle$, where C is a *ternary* relation of "similarity": $C_x yz$ for "y is more similar (closer) to x than z is."

The key clause in the truth definition now reads:

$$M \models \varphi \Rightarrow \psi \, [w] \quad \text{if} \quad \psi \text{ holds in all those } \varphi\text{-worlds which are closest to } w.$$

Thus again, minimal deviation is the guiding idea. The picture here is that, if φ is true at w, $\Rightarrow$ will reduce to ordinary $\rightarrow$ (assuming w to be closest to itself):

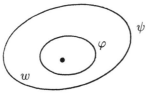

indicative.

But otherwise, the picture may be something like this:

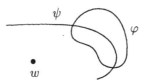

counterfactual.

There are some minimal structural assumptions on C here: for a fixed vantage point x, $\lambda yz \, . \, C_x yz$ is to be *transitive* and *irreflexive*. The minimal logic of this scheme may be axiomatized as follows (see J. Burgess, Quick Completeness Proofs for Some Logics of Conditionals, *Notre Dame Journal of Formal Logic* 22, 1981, pp. 76–84):

a. $A \Rightarrow A$,

b. $A \Rightarrow B \, / \, A \Rightarrow B \vee C$ ("Weakening of the Consequent"),

c. $A \Rightarrow B, A \Rightarrow C \,/\, A \Rightarrow B \wedge C$,

d. $A \Rightarrow B, C \Rightarrow B \,/\, A \vee C \Rightarrow B$,

e. $A \Rightarrow B \wedge C \,/\, A \wedge B \Rightarrow C$.

The reader may find it useful to check these principles. Typically invalid will be principles such as Strengthening of the Antecedent ("monotonicity") or Transitivity. (Lewis presents various intuitive counterexamples.)

As usual, further constraints on our models correspond to enforcing validity of additional principles. Especially in conditional logic, where intuitions have clashed repeatedly concerning the "validity" of certain inference schematic, such as semantic perspective can make the options more perspicuous.

One obvious further condition justifies our picturing the C_x-order as a nest of concentric circles:

$$\forall xyzu\colon C_x yz \to C_x ys \vee C_x sz \qquad (almost\text{-}connectedness).$$

(Equivalently: "$\lambda yz \,.\, \neg C_x yz$ is transitive.") The corresponding additional axiom is rather cumbersome. Even more restrictive is the constraint corresponding to a principle of conditional reasoning which Stalnaker endorsed (and Lewis rejected):

$$\varphi \Rightarrow \psi \ \vee \ \varphi \Rightarrow \neg\psi \qquad (\text{Conditional Excluded Middle}).$$

Its semantic effect is to make the alternatives a *linear order*, with a unique closest φ-world. All these conditions are still "horizontal," in the sense that no iteration principles for $\Rightarrow$ are validated, or semantically, that no "index principles" are generated connecting the similarity order as viewed from different vantage points x. A geometrical example of the latter type would be

$$\forall xyz\colon (C_y xz \wedge C_z yx) \to C_x yz \qquad (\text{Triangle Inequality}).$$

Evidently, the resources of this model theory largely remain to be explored.

The latter conclusion also holds for axiomatizations of conditional logics, and indeed the whole logical theory of this area. (One problem is that the Henkin model technique, which has proved so powerful in modal logic, meets some curious difficulties in this area, having to do with a failure of compactness. For a principled exploration, see F. Veltman, 1985, *Logics for Conditionals*, dissertation, Filosofisch Instituut, University of Amsterdam.)

The Lewis semantics has some peculiar features, which come out when the above idea is extended to the case of *infinite* sets of worlds W. In that realm, there need not be "closest" φ-worlds (there could be ever closer worlds where you are taller than you actually are), and the above truth clause requires amendments—none of them wholly satisfactory. This possible diversity of truth definitions (in addition to the diversity in constraints on model classes which we have encountered already) is an interesting semantic phenomenon, challenging the surely not self-evident presupposition of much recent research that unique truth conditions must be forthcoming for linguistic expressions (rather than some "range of explication"). Nevertheless, this diversity also reflects a certain diffi-culty in the notion of "similarity," which seems to lack a clear intuitive basis to many observers. Thus, it is no surprise that rival semantic accounts have been produced in the seventies.

Premise Semantics

One interesting alternative, due to Kratzer and Veltman, forms, in a sense, a compromise between "syntactic" and "semantic" views. Models are now triples $\langle W, P, N \rangle$ where the function P assigns, to each world $w \in W$, the *premises* (or prejudices) reigning there—or rather, their semantic correlates: $P(w)$ is a set of subsets of W. One can think of $P(w)$ as my prejudices, in state w, about what the world is like. (Notice that these may be wrong, in the sense that w fails to belong to some of these.)

The key semantic clause now becomes the rather formidable equiv-alence $M \models \varphi \Rightarrow \psi[w]$ if every set $S \subseteq P(w)$ which is consistent with the extension of φ (the set of worlds where φ is true)—that is, $\cap S \cap \mathrm{ext}(\varphi) \neq \emptyset$—can be extended to a set $S^+ \subseteq P(W)$ consistent with φ such that S^+ with φ implies ψ—that is, $\cap S^+ \cap \mathrm{ext}(\varphi) \subseteq \mathrm{ext}(\psi)$.

With some reformulation, the earlier Ramsey pattern reemerges.

As it turns out, premise semantics and Lewis semantics are intimately related. There is a trade-off between C-flexibility and P-flexibility, to the extent that models can be mutually translated. Thus, the minimal logic is the same in both cases (a result due to Lewis and Veltman, independently. See the *Journal of Philosophical Logic*, vol. 10, 1981— an issue on conditionals—for details). Such unpremeditated equivalence results, and there are more examples, at least establish a certain "sta-bility," and hence respectability, for the basic conditional logic—as was the case with similar discoveries concerning the logic of intuitionism.

There are also quite different approaches to conditional semantics, notably the *probabilistic* one of E. Adams and R. Stalnaker, who have tried to relate probabilities for conditionals to conditional probabilities. These fall outside of the scope of these notes.

The logic of conditionals has not at all been exhausted by the above discussion. For instance, conditionals also play an important role at a *contextual* level, in setting up or modifying *contexts* ("*if A, ...*"). Thus, they provide also a good case for a general theory of "contextual dynamics," a conspicuous trend these days. (A first influential statement has been R. Stalnaker, 1972, Pragmatics, in D. Davidson and G. Harman, eds., *Semantics of Natural Language*, Reidel, Dordrecht, 380–397.) In a sense, understanding conditionals involves understanding most of the central aspects of reasoning.

4 Combinations

Literature

Relevant parts of

Chellas, B. 1980. *Modal Logic: An Introduction.* Cambridge, Mass.: Cambridge University Press.

Burgess, J. 1979. Logic and Time. *Journal of Symbolic Logic* 44, 566–582.

Van Eck, J. 1981. *A System of Temporally Relative Modal and Deontic Logic, with philosophical applications.* Ph.D. dissertation, Rijksuniversiteit, Groningen. Also in *Logique et Analyse*, 1983.

Thomason, R. 1984. Combinations of Tense and Modality. In D. Gabbay and F. Guenther, eds., *Handbook of Philosophical Logic*, vol. II, Dordrecht: Reidel, 135–165.

Time and Modality

The most fundamental combination in the tradition has temporal and modal elements intertwined. In an *analytic* approach, these are treated as separate components:

$W \times T$-models have a bundle of worlds, sharing a common time pattern. (More exotic possibilities, with world-dependent time structure, are conceivable.) Evaluation takes place at an "index" with two components:

$$M \models \varphi[w, t].$$

This is only the beginning of a long road in intensional semantics, where the index carries more and more contextual parameters. Two key clauses exemplify the spirit of this perspective:

$$M \models F\varphi\,[w, t] \quad \text{if} \quad \exists\, t' > t \colon M \models \varphi[w, t'],$$

31

$M \models \Box\varphi\,[w,t]$ if   (future in the same world)
(option 1, Montague) $M \models \varphi[w',t']$ for all w',t' (strong necessity)
(option 2, others) $M \models \varphi[w',t]$ for all $w'R_t w$ (temporal necessity).

In the latter case, one has a *family* of accessibility relations R_t, one at
each moment in time, allowing for changes in access. One common choice
here is to let R_t imply sharing the same past up till (and including) t,
and the familiar picture emerges of a branching tree of possible world
courses

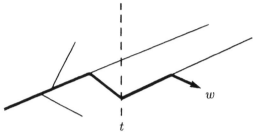

This is a prevalent conception, both in science and literature (witness
Borges' famous story "The Garden of Forking Paths").

In a "standard model" for this tradition, one might even *define* the
worlds as all functions from points in time to valuations, with accessibility as explained above. The logic of this scheme has not been determined
yet; but even so, this semantics has proved its utility.

Applications: Determinism

The earlier Sea Battle argument had the following structure, when read
with temporal necessity:

$$\Box_t(p_{t+1} \to q_{t+2}), \quad \Box_t(\neg p_{t+1} \to \neg q_{t+2}), \quad \Box_t(p_{t+1} \vee \neg p_{t+1}),$$
$$\text{and yet } not\colon \Box_t\, q_{t+2} \vee \Box_t \neg q_{t+2}.$$

In R. Taylor's case for fatalism, a purported "symmetric" version is
presented with respect to the past: the admiral's acts now being to read
(or not read) a certain newspaper headline about a battle (or no battle)
yesterday. Taylor claims that everyone finds this past version compelling,
and hence that its future version must also be accepted. But, on the
above semantics, the past version looks like this:

$$\Box_t(p_{t+1} \to q_{t-1}), \quad \Box_t(\neg p_{t+1} \to \neg q_{t-1}), \quad \Box_t(p_{t+1} \vee \neg p_{t+1}) \text{ are true,}$$
$$\text{and so is } \Box_t q_{t-1} \vee \Box_t \neg q_{t-1}.$$

But, this is no case of implication: the latter statement is valid anyway,
as our semantics incorporates *necessity of the past* ("irrevocability").

Other consequences of the semantics force one to be careful with common logical ploys. For instance, Aquinas offered the following refutation

of the traditional problem of reconciling human free will (and hence responsibility, and sin) with divine omniscience. (In the Middle Ages, these issues were burning matters.) The problem runs like this. If I am sitting here, then God saw me sit, at the moment of Creation. Now, whatever God saw then, necessarily takes place. Hence, my sitting here is necessary: it could not have been otherwise. Aquinas notes that the necessity here could be "de re" (if φ, then necessarily ψ) or "de dicto" (necessarily, if φ, then ψ). As in the Sea Battle case, the latter obtains, not the former—and the conclusion fails. But, on the above semantics, this scope distinction is *spurious* when applied to necessity of past events: $\Box_t(p_{t-1} \to q_{t+1})$ is equivalent to $p_{t-1} \to \Box_t q_{t+1}$. Hence, one will have to analyze the premises more carefully, in order to reject the conclusion.

Here is also the place for an analysis of Diodorus' Master Argument.

Version 1. With a purely metaphysical reading of $\Diamond$ and $\Box$, accessibility between worlds will be an unanalyzed relation R in addition to the temporal order $<$. The premises of the Argument may then be viewed as constraining the temporal order (as was explained in Chap. 1), as well as enforcing a certain interaction with the metaphysical pattern:

Axiom 4 corresponds to the condition that

$$\forall x : \forall y(y < x \to \forall z(Rxz \to y < z)).$$

But then, the following observation will produce a reduction from modality to temporality:

$$\Diamond\varphi \to (P\varphi \vee \varphi \vee F\varphi).$$

For, suppose that $\Diamond\varphi$ is true at a world x, and hence φ true at some alternative y with Rxy. By axiom 3, x has some temporal predecessor $z < x$. The above interaction principle then implies $z < y$ as well. By axiom 2, x, y must be temporally comparable, and the conclusion follows.

Version 2. With the semantics of temporal modality, however, the three purely temporal premises of the Master Argument are true, as is Necessity of the Past. Nevertheless, the conclusion might obviously be false: facts occurring in some side-branch of our world may create possibility without actual occurrence anywhere in our actual history.

Holistic Approaches

In another perspective, tense and modality may be just two aspects of what is essentially one notion. Models may then be regarded as sets of possible *world-states*, ordered by *possible succession* (and, perhaps, simultaneity). Time, or possible worlds then arise through an act of abstraction: the latter may be thought of as the *maximal chains* through this structure, the former as the (linear) order within these chains.

Formally, one has models $\langle S, <, V \rangle$, where V assigns truth values to atomic propositions at states $s \in S$. Various formal requirements on S come to mind, both elementary ($<$ is to be irreflexive and transitive) and higher-order ($\langle S, < \rangle$ might be a tree; the order pattern might be required to be the same in all maximal chains). In this more tentative section, we shall not be more specific on these issues.

The truth definition explicates truth of formulas at world/state pairs, with key clauses, say,

$$M \models p[w, s] \quad \text{iff} \quad s \in V(p) \quad \text{(the context } w \text{ plays no role)},$$
$$M \models F\varphi[w, s] \quad \text{iff} \quad \text{for some } s' > s: \; s' \in w \text{ and } M \models \varphi[w, s'],$$
$$M \models \Box\varphi[w, s] \quad \text{iff} \quad \text{for all } w', \text{ if } w'|s = w|s, \text{ then } M \models \varphi[w', s]$$
$$\text{(where } w|s \text{ is the world-course } w \text{ up to}$$
$$\text{and including } s\text{).}$$

Again, this will reflect the "Occamist" position that the past is necessary ($Pq \rightarrow \Box Pq$ being valid) while the future is not ($Fq \rightarrow \Box Fq$ being invalid).

The logic of this scheme has no surprises in its separate components (tense and modality by themselves). The main interest resides in their *interactions*, such as the above-mentioned validity $Pq \rightarrow \Box Pq$, or the future principles

$$\Box F \Box F \varphi \rightarrow \Box F \varphi,$$
$$\Diamond F \Diamond F \varphi \rightarrow \Diamond F \varphi.$$

Notice that $\Diamond F$ expresses a weak modal "will," where $\Box F$ has a strong sense of "inescapability":

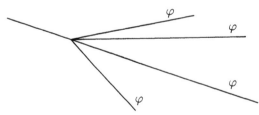

Both these meanings have been put forward for the natural language future tense. The $\Diamond F$-principle has the technical interest of presupposing "piecing together" of branches into one. Stronger "axioms of choice" in this spirit are available too. As a matter of interest, various defective completeness proofs have been announced for their models of "branching time," so there must be some hidden subtleties.

The above logic lacks *substitutivity*. For instance, although $Pq \rightarrow \Box Pq$ was valid, its substitution instance $PFq \rightarrow \Box PFq$ is not. This feature answers a medieval worry about Occam's view: that the necessity of the past could be logically "transferred" to the future. John Burgess has proposed related semantics, however, where substitutivity is restored:

one basic difference being that, on his account, truth of atomic statements can be dependent upon both state and world (which invalidates $Pq \rightarrow \Box Pq$). (Steve Thomason has conjectured that the Burgess logic is the "substitution-closed core" of the earlier system.) Many other semantic variants exist.

Other Combinations

Similar combined semantic pictures can be used to obtain temporal variants of the earlier conditional semantics. *Tensed conditionals* are a basic fact of life, for instance in practical deontic logic, where the whole game consists in ever changing conditional obligations.

Temporalizing earlier semantics is not the only intensional combination of interest. A quite different example is the interplay of *knowledge* and *action* exhibited in a combined possible worlds semantics for epistemic logic and dynamic logic. For instance, it may be studied how actions affect epistemic alternatives, with corresponding "interchange inferences" for knowledge and action result operators. Or trade-offs may be studied between lack of knowledge and genuine indeterminism of actions. Such themes are developed in R. Moore, *A Formal Theory of Knowledge and Action*, Technical Note 320, Artificial Intelligence Center, SRI International, Menlo Park, 1984.

We shall conclude with a topic illustrating a quite different, scientific atmosphere of thinking about time and modality.

Time, Space and Causality

In the philosophy of space and time, there has been a recurrent program called the Causal Theory, found with Leibniz, and revived in this century by A. Robb. The main idea here is that our primary physical entities are *possible events*, ordered by *possible causal precedence*. Out of these, space and time are to be constructed, with the usual properties. A very good exposition is J. Winnie, The Causal Theory of Space-Time, in J. Earman et al., eds., *Foundations of Space-Time Theories*, University of Minnesota Press, Minneapolis, 1977, 134–205.

In Leibniz' original attempt, the relation $<$ of possible precedence was assumed to be transitive, irreflexive and almost-connected. (Interestingly, these are also Lewis' original postulates on similarity.) Simultaneity of events was then defined as mutual non-precedence, with "time" becoming the order of the simultaneity classes, "space" the ordering pattern inside these. Unfortunately, as Winnie shows, there is not enough structure left in the latter to define non-trivial geometric relations (such as betweenness). Thus, the Causal Theory fails for classical physics.

With *relativistic* physics, the program turns out to fare much better: one of many indications that the latter is conceptually more elegant. One may think of the familiar light cones in Minkowski space:

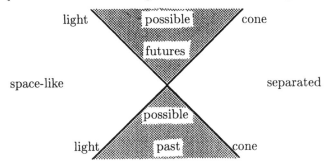

Here, < becomes the relation between an event and all events in the interior of its forward light cone. As it turns out, the first-order logic of this structure does contain both its "chronometry" and "geometry," while its second order theory even adds the topological and metric structure. (There is a representation theory proving all this, with strong resemblances to Tarski's program for elementary geometry.)

One basic feature of this semantics it that (possible) *precedence* between events is spatially restricted: one only calls e (potentially) "before" e' if some meaningful causal chain of events could lead from e to e'. This non-classical feature seems highly reasonable, also in natural language—and one can see how relativistic concerns could enter semantics even without worrying about the effects of travel near the speed of light.

It remains unknown what is the tense logic (or modal tense logic) of the above structure. A partial result is found in R. Goldblatt, Diodorean Modality in Minkowski Space-Time, *Studia Logica* 39, 1980, 219–236. The modal logic of $\leq$ is the system *S4.2* (see Chap. 2), with axioms

$$\Box\varphi \to \varphi, \quad \Box\varphi \to \Box\Box\varphi, \quad \Diamond\Box\varphi \to \Box\Diamond\varphi.$$

Goldblatt's proof follows a very interesting conceptual road, starting from logical Henkin models, and ending up with the Minkowski model of actual scientific practice. Such bridges between our kind of formal semantics and the models of science are regrettably scarce. Jens-Erik Fenstad has suggested various interesting analogies between the semantics of this chapter and scientific uses of parallel worlds that might be pursued. One such program was proposed already earlier in the seventies by Albert Dragalin at Moscow, concerning axiomatizations for various physical modalities in space-time. (See V. B. Shehtman, Modal Logics of Domains on the Real Plane, *Studia Logica* 42:1, 1983, 63–80; which also documents its author's prior discovery of Goldblatt's result in 1976.)

II Recent Developments 1: Computational Applications

The main emphasis in the preceding chapters has been on the original motivations from philosophy and linguistics guiding the development of Intensional Logic. In addition to these, however, other spheres of action have opened up for the enterprise. Some of these lie in the heartland of mathematical logic, others have only arisen with the advent of computer science. The purpose of the present part is to illustrate the nature and diversity of these newer applications.

5 Mathematics and Computer Science

Mathematics and Foundations

Although mathematics is often presented as a paradigm of "extensional" thinking, there have been dissident views. Intuitionists have always stressed the differences between mathematical notions and their usual set-theoretic, extensional reductions. For instance, a function viewed as a rule is more than a range of ordered pairs. Thus, *intuitionism*, and constructive mathematics generally, is a natural field for Intensional Logic. And indeed, at a very early stage, connections were established between intuitionistic logic and modal logic by Kurt Gödel. In the sixties, Saul Kripke then exploited Gödel's embedding of intuitionistic logic into the modal logic *S4* to make possible worlds semantics serve intuitionistic logic too. (See Chap. 7 for elaboration. A comprehensive recent survey is A. Troelstra and D. van Dalen, 1988, *Principles of Constructive Mathematics*, vol. I and II, North-Holland, Amsterdam, to appear.)

Intuitionistic logic is by no means the only way of accounting for intensional aspects of mathematical reasoning. An alternative, using explicit knowledge operators from epistemic logic may be found in S. Shapiro, ed., 1985, *Intensional Mathematics*, North-Holland, Amsterdam.

Knowledge and proof are often intertwined in mathematics. As has been mentioned already in Chapter 2, modal logic has also been applied extensively to so-called *provability logic* of mathematical theories. This reflects Hilbert's original intuition that, even if mathematical theory itself is about some complex domain, its *metatheory* may be much more easily understood, especially the general laws governing its proof structure. Modal formalisms often provide a medium for stating such laws (such as Löb's Theorem) which is reasonably expressive, and yet simple enough to admit of complete axiomatization. Actually, this enterprise is by no means finished. For instance, current research in The Netherlands (de Jongh/Smoryński/Veltman/Visser) is concerned with "modalizing" other important meta-notions, such as *relative interpretability* between various arithmetical theories.

The latter type of research can be explained more easily by reference to a general setting of arbitrary (not necessarily arithmetical) first-order theories. Let

$$A \triangleright B$$

mean that theory A (finitely axiomatized by A, that is) can interpret B in the usual Tarskian sense. What is the general logic of this operator? Here again, an analogy from Intensional Logic is helpful. The wedge $\triangleright$ resembles the *conditionals* of Chapter 3 in that it may be shown to obey the following laws:

a. $A \Rightarrow A$,

b. $A \Rightarrow B \ / \ A \Rightarrow B \vee C$,

d. $A \Rightarrow B, C \Rightarrow B \ / \ A \vee C \Rightarrow B$,

e. $A \Rightarrow B \ / \ A \wedge C \Rightarrow B$ (Strengthening the Antecedent).

(A principle which fails, however, is Conjunction of Consequents:

c. $A \Rightarrow B, A \Rightarrow C \ / \ A \Rightarrow B \wedge C$.

Compare $A = p$, $B = p$, $C = \neg p$.)

Question: Do these four principles axiomatize the complete propositional logic of relative interpretability between arbitrary first-order theories?

It is actually easy to give a complete possible worlds semantics for the above logic. Consider structures for $S5$, consisting of equivalence classes of possible worlds. Now, call a formula $A \triangleright B$ *true* in such a model if every world verifying A has some equivalent world verifying B. This truth condition validates the above inferences, and in fact (as may be proved quite simply) essentially *only* those. To answer the above question, then, it would have to be shown that counterexamples on $S5$-models can be "encoded" via suitable substitutions of concrete first-order sentences, reading $\triangleright$ as actual interpretability.

Again, even the latter uses of Intensional Logic do not exhaust its mathematical scope. Modal structures also arise naturally in ordinary, non-foundational settings (witness, e.g., the so-called *topological interpretation* of intuitionistic logic, or the category-theoretic treatment of *geometric modality* in R. Goldblatt, 1979, *Topoi*, North-Holland, Amsterdam).

It is not just the foundations of mathematics where Intensional Logic has penetrated. The same holds for the foundations of *physics*; witness such examples as the intensional semantics for Quantum Logic (see R. Goldblatt, Semantic Analysis of Orthologic, *Journal of Philosophical Logic* 1, 1971, 91–107). Also, modal elements have been introduced in the very set-up of classical physical theories, such as Mechanics (see A. Bressan, 1972, *A General Interpreted Modal Calculus*, Yale University Press, New Haven and London).

In many of these cases, the intensional element has to do with some general methodological feature of empirical science. As illustrations, one may think of modal and epistemic aspects of *probability* (see Lenzen's book cited in Chap. 2), or *causality* and *natural laws* (see, e.g., W. Salmon, Laws, Modalities and Counterfactuals, *Synthese* 35, 1977, 191–229). Another example, concerning conditionals and *verisimilitude*, will be found in Chapter 10.

But certainly the most exciting new developments of Intensional Logic have taken place in various areas of a new discipline, viz.

Computer Science

From its inception, computer science has been intertwined with logic in various ways. Computability is connected with notions and results from Recursion Theory; but also Model Theory plays an important role in the *semantics of programs*. In fact, existing logical systems have even served as models for setting up programming languages, such as LISP or PROLOG. Many of these connections have to do with standard logic. But gradually, Intensional Logic is becoming more prominent too, across a wide range of activities, such as proving correctness for programs or protocols in computer science proper, and in setting up knowledge representation in Artificial Intelligence. More concrete examples will be found below, in a survey organized under roughly the same headings as the previous chapters.

Time and Temporal Representation

One of the earliest uses of temporal logic in computer science had to do with proving statements about the behavior of *concurrent programs*. A number of processors is to execute various tasks in parallel, and we want to prove various statements about the process. These include not only *correctness* of the final result computed, but also certain desired *behavior* on the way toward achieving that goal. For instance, execution should be *fair*, in that any processor who is able to function at points throughout the future of some stage in the process, must in fact be activated. A formula expressing this property is

$$GFq \rightarrow \neg Gp$$

where p expresses the processor's being in some state, and q the fulfillment of some exit condition for that state. Another example is absence of *deadlock*: a situation where all processors have come simultaneously in states from which they cannot exit. Again, this is easily expressed in temporal logic. (A ground-breaking paper in this area is A. Pnueli, The Temporal Semantics of Concurrent Programs, *Theoretical Computer Science* 13, 1981, 45–60.)

Not just classical temporal logic has been used in this setting. For instance, newer temporal representations in terms of intervals and *events* (to be described in Chap. 6) have also been used for these purposes, witness L. Lamport, 1985, *Interprocess Communication. Final Report*, SRI International, Menlo Park. The reason for this change is that concurrent processes invite us to rethink the earlier semantic framework of Chapter 1. (In fact, finding a good conceptual framework for describing parallel processes is one of the most urgent issues in computation today.) It may be both conceptually and computationally more perspicuous to model parallelism as a partial order of events, which can precede or overlap each other. And of course, such a change in perspective would also have important consequences for specification and design. Technical research emanating from the latter motivation has concentrated on finding good sets of primitive relations between events, generating well-behaved logical theories. (See P. Ladkin, 1987, Models of Axioms for Time Intervals, Kestrel Institute, Palo Alto; P. Ladkin and R. Maddux, 1987, The Algebra of Convex Time Intervals, Kestrel Institute, Palo Alto/ Department of Mathematics, Iowa State University at Ames.)

The move from classical time lines to event structures has also been made in Artificial Intelligence. The latter concept seems closer to common sense representations of time, which presumably guide our own practical thinking. Therefore, such issues also become important in setting up automated systems for such tasks as planning a sequence of actions toward a certain goal. (An influential paper in this vein is J. Allen, 1983, Maintaining Knowledge about Temporal Intervals, *Communications of the Association for Computing Machinery* 26, 832–843. Other references will be given in Chap. 6 below.)

We conclude with a more "classical" example, being an *interval tense logic* developed by J. Halpern and Y. Shoham, 1986, A Propositional Modal Logic of Time Intervals, *Proceedings Symposium on Logic in Computer Science*, IEEE, Boston. (See also Y. Shoham, 1986, *Reasoning about Change: Time and Causation from the Standpoint of Artificial Intelligence*, dissertation, Department of Computer Science, Yale University.) This logic is a fair example of the nature of application of Intensional Logic in computer science: existing basic ideas remain useful, but they have to be adapted in non-routine, and often very interesting ways.

Let us consider convex intervals on strict linear orders, given by their endpoints. Propositions can now be true or false at such intervals—the earlier case of points in time being included via unit intervals. Important operators again express relative positions of events:

a. BEGINφ is true at (t_1, t_2) if there exists $t_3 < t_2$ such that φ is true at (t_1, t_3)

(The notation (t_1, t_2) always presupposes that $t_1 \leq t_2$.)

b. STARTφ is true at (t_1, t_2) if there exists $t_3 > t_2$ such that φ is true at (t_1, t_3)

c. BEFOREφ is true at (t_1, t_2) if there exists $t_3 \leq t_1$ such that φ is true at (t_3, t_1)

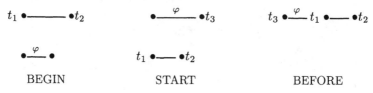

BEGIN	START	BEFORE

As in Chapter 1, one can study expressive power, or possible axiomatizations for this logic, and its various extensions. (There is this technical difference that validity now involves quantification over *binary*, rather than *unary* sets of points: with a corresponding upward leap in complexity.)

Here we merely point out an interesting *reinterpretation* of this logic (noted by Yde Venema) which uses a trick from *The Logic of Time* (cf. Chap. 1). Intervals (t_1, t_2) may be represented as *points* in the cartesian product $T \times T$ of the underlying temporal structure T with itself. Then, the above operators become rather *topological* ones, referring to positions along the natural dimensions of movement in the plane:

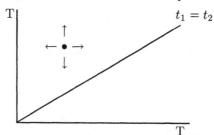

As may be checked quite easily, BEGIN refers to moving southward, START goes north, and BEFORE goes west. (Of course, moving east also has a counterpart with Halpern and Shoham, who present more operators than those shown here.) Thus, Logic of Time may also be Logic of Space.

Epistemic Logic

As has been observed, despite their different background, modal logic
and epistemic logic share quite similar formalisms. The pioneering work
here was J. Hintikka, 1962, *Knowledge and Belief*, Cornell University
Press, Ithaca. As Hintikka observed, various basic logics of knowledge
and belief exhibit an *S4*- or an *S5*- like structure, with corresponding
possible worlds (or "epistemic states") semantics. For instance, a basic
modal axiom such as Distributivity

$$\Box(\varphi \rightarrow \psi) \rightarrow (\Box\varphi \rightarrow \Box\psi)$$

now expresses that known consequences of knowledge are themselves
known. While this "closure axiom" has generated a lot of controversy
(actual knowledge being less than closed under deductive consequence),
it is quite plausible on an account of *ascribed* knowledge. And the latter
notion is rather natural in computer science, where an "anthropomor-
phic" view of processors, ascribing knowledge to them as if they were
humans, has turned out quite useful in reasoning about computation.
("As soon as processor i knows that its neighbour j knows the answer
to x, it will ask j to ....")

Here are some other *S5*-axioms, with their epistemic readings (whose
acceptability is left to the reader's judgement):

$$\Box\varphi \rightarrow \varphi \qquad \text{what is known is true;}$$
$$\Box\varphi \rightarrow \Box\Box\varphi \qquad \text{Positive Introspection:}$$

if something is known, this fact itself is known;

$$\neg\Box\varphi \rightarrow \Box\neg\Box\varphi \qquad \text{Negative Introspection:}$$

if something is not known, this ignorance itself is known.

An important feature of Hintikka's logic is the possibility of indexing
knowledge operators, so that $\Box_i\varphi$ expresses person/processor i's knowledge that φ. In the corresponding epistemic structures, each i may have
a different accessibility relation on the worlds, encoding her views of reality. For instance, in the following structure, person 1 knows at world
x that p, person 2 does not know p, but she knows that person 1 knows
that p:

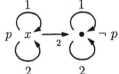

World x verifies: $\Box_1 p$, $\neg\Box_2 p$, $\Box_2(\Box_1 p \vee \Box_1\neg p)$.

Evidently, this is also the proper generality for arguing about systems
of processors.

In Hintikka's own work, this logic was used to clarify issues in the Theory of Knowledge; as it enabled him to draw subtle distinctions as to assertoric force of epistemic statements. In fact, this issue is a pervasive one, also in ordinary semantics of natural language. Individual speech acts and dialogue in general have an obvious cognitive dynamics, having to do with various modes of commitment. W. Lenzen, in his book cited in Chapter 2, investigates such additional operators as "überzeugung" (conviction), whose force should be in between belief and knowledge. Again, there is a very practical computational interest in these matters. Current natural language processing systems are moving toward system/user dialogues, whose proper understanding presupposes a theory of assertoric force. (The system needs to "know" what the user knows, conjectures, etcetera, to give answers which would be considered appropriate in ordinary discourse.) One survey of some relevant problems is E. Thijsse, 1987, *Kripke Models for Knowledge Bases I: S5-Miniatures*, Department of Language and Informatics, University of Tilburg. Incidentally, Thijsse proposes the following standard assertoric force for ordinary indicative statements:

$$BK\varphi: \quad \text{I believe that I know } \varphi.$$

With the above, we have seen two major motives for using epistemic logic in a computational setting. Especially the first has sparked off an impressive body of work, much of it emanating from the IBM Almaden Research Center at San Jose, California. For a good survey, as well as state-of-the-art papers, see J. Halpern, ed., 1986, *Theoretical Aspects of Reasoning about Knowledge*, Morgan Kaufmann Publishers, Los Altos. Of the many topics arising in this way, we mention a few.

Originally, Joe Halpern, Moshe Vardi and Ron Fagin tried to model the iteration involved in i's knowing that j does not know if k knows ... in terms of inductively defined "knowledge structures" of ever greater depth. But here, possible worlds semantics proved its mettle. Although independently motivated, knowledge structures turned out to reduce to the original Hintikka possible worlds models. Thus, the classical framework still stands, with "worlds" viewed as global states of a system of distributed processors (i.e., vectors of local states for each processor). From each global state, processor i will see those other global states as indistinguishable in which she has the same local state. This makes accessibility an equivalence relation, and hence the basic epistemic logic here is $S5$.

As was observed before, epistemic reasoning involves *ignorance* as much as *knowledge*. Many logical questions surround this phenomenon, as temporary conclusions from ignorance may have to be withdrawn when positive knowledge grows. For an analysis of "stable" epistemic situations, see R. Moore, 1983, *Semantical Considerations on Non-Monotonic Logic*, SRI International, Menlo Park, as well as J. Halpern and

Y. Moses, 1984, Towards a Theory of Knowledge and Ignorance, in *Proceedings of the Workshop on Non-Monotonic Reasoning*, AAAI, 1984.

Another prominent topic in the current literature again reflects an older philosophical concern. Once individuals are gathered into groups, various forms of *collective knowledge* come into play. For instance, note the practical difference in a room filled with spies who each know where the secret document is hidden, but no more, and then a similar situation where each spy knows that the others know that location too. In fact, it has been argued that many forms of successful collective action, whether on an adventure trip or merely in an ordinary dialogue, presuppose *common knowledge* of certain basic principles on the part of participants:

$$\text{all combinations } \Box_i \Box_j \Box_k \ldots \varphi \text{ are required.}$$

This is really a new notion, with an infinitary flavour, satisfying such *fixed-point axioms* as

$$C\varphi \leftrightarrow \varphi \wedge EC\varphi,$$

where "C" denotes common knowledge, and "$E\psi$" stands for "everybody knows that ψ," i.e., the conjunction $\Box_1\psi \wedge \Box_2\psi \wedge \ldots$.

A perceptive early study of this notion is D. Lewis, 1969, *Convention. A Philosophical Study*, Harvard University Press. In the meantime, some sophisticated logical techniques have been applied to the logic of C, including modal logics enriched with so-called fixed-point operators. As it has turned out, there are interesting connections here with current work in economics on agreement between rational agents. (For a dissident view, advocating a major conceptual overhaul in order to account smoothly for the circularity or regress inherent in common knowledge, see J. Barwise, 1988, The Situation in Logic IV: On the Model Theory of Common Knowledge, Report CSLI–88–122, CSLI, Stanford.)

One of the most fascinating topics in this area right now is precisely the *interaction* between epistemic and other intensional notions, such as flow of *time* or purposeful *communication*. A standard example in the area, which illustrates some basic issues, is Byzantine Agreement. Two Byzantine generals are encamped on opposite hill tops. Tomorrow they might attack the enemy, located down in the valley—but this will be successful only if *both* of them act together. So, they need to synchronize departure times. The only means of communication is to send messengers through contested territory, who might be killed in transit. (In technical terms, this is not a perfect communication channel: the usual situation in hardware.) Now, suppose general 1 gets a message through to general 2, announcing his intention to attack tomorrow at 6 AM, and suppose also that general 2 wants to so the same. Of course, he has to *confirm* this to general 1. But also, he needs to know if his confirmation got through; which requires one more message from general 1, which again requires ... : an infinite regress is under way, and no attack seems possible. The generals' problem is that they need common knowledge of their intentions, and they cannot get it with defective means of communication. A sparkling paper on this, and similar problems, as well as some practical remedies, is J. Halpern and Y. Moses, 1984, Knowledge and Common Knowledge in a Distributed Environment, *Proceedings of the Third ACM Conference on Principles of Distributed Computing*, 1984, 50–61. Both the philosophical and the mathematical situation are rather subtle here (combination of the separate logics for components often gives jumps in complexity of valid inference).

Especially, the latter theme of *complexity* is characteristic for this area. We do not just want plausible logics of knowledge, but also logics whose complexity still leaves us some hope that human inference might actually work that way. This additional concern is a very welcome enrichment of the somewhat placid waters of epistemic logic as it used to run its course in the philosophical community. In any case, the issue of complexity and computational plausibility is a large one, which deserves much deeper discussion. As it is largely orthogonal to Intensional Logic per se, however, we shall not pursue this theme here. We conclude by pointing out another moral of the above. Actual knowledge is knowledge in action, in an environment of time and communication. If we are to really understand this phenomenon, we will have to go beyond the usual "piecemeal strategy": the interaction of various intensional logics in this setting may even be more interesting than the individual logics by themselves.

Conditional Logic

Classical conditional logic has not remained unaffected by computational developments either. For instance, when read "dynamically," sets of conditionals become systems of rules, which can be used in automated systems of *non-monotonic* reasoning. (See D. Nute, 1986, *A Non-Monotonic Logic based on Conditional Logic*, report 01–0007, Advanced Computational Methods Center, University of Georgia, Athens, GA.) Indeed, the syntax and semantics of conditional logic provide a good model for studying basic aspects of non-monotonicity in computer science. For instance, in John McCarthy's famous notion of *circumscription*, the informative content of a set of premises Σ is taken to be not just what Σ asserts to exist, but also: what it fails to mention. Technically, one thus considers, not *all* models for Σ, but only those which are *minimal* in some sense. Within this smaller class, Σ may have many more consequences than just its logical ones. (See J. McCarthy, 1980, Circumscription—A Form of Non-Monotonic Reasoning, *Artificial Intelligence* 13, 295–323.) For a concrete example, think of minimality of extensions for predicates in models of Σ. E.g., (D, R_1, R_2) is a minimal model for $\varphi(X_1,X_2)$ if no further decrease of R_1 or R_2 still leaves a model for φ. (Note the connection with so-called *Pareto optimality* in economics.) This will make, say, $\forall x Sx$ a minimal consequence of one individual premise Sa. But, this conclusion may fall upon acquisition of further information (such as $\neg Sb$). Minimal models for φ,ψ need no longer be minimal models for φ alone—and hence we have again a form of non-monotonicity.

Now, conditional logic does two things for us here. First, its semantics provides a general abstract scheme behind all kinds of minimality proposed in the literature on circumscription. (Cf. the earlier-mentioned dissertation Shoham 1986, which eventually arrives at this abstraction in the end.) Moreover, its axiomatics shows what general logic governs the notion of minimal consequence, as opposed to classical consequence. Although one loses certain familiar principles, such as Monotonicity or Transitivity, others will remain, as summed up in the basic conditional logic of Chapter 3. (The validity of axiom 5 actually needs one additional assumption, viz. that the ordering over which one minimizes is *well-founded.*) As has been pointed out by David Lewis already, the new situation often brings more subtle substitutes for lost validities. E.g., instead of Transitivity, we still have

$$A \Rightarrow B, \qquad A \wedge B \Rightarrow C \quad / \quad A \Rightarrow C.$$

Another way of viewing the above developments is, of course, that, like its colleagues in Intensional Logic, Conditional Logic has suddenly acquired a wealth of concrete, computational models.

Digression. Conditional logic is not the only "classical theory" involved in the phenomenon of *minimal modelling*, which is now so widespread in computer science. (Circumscription provided one example; but so does Logic Programming, or the theory of Abstract Data Types.) The general situation also contains a natural *modality*. Given a set of statements T, we have the class of all its models, needed for determining the standard logical consequences of T. But, there will also be some binary relation R on models, such that we can consider only the R-minimal models for T, and see what follows there. More abstractly, for any class of models A,

$$\mu(A) = \{M \in A \mid \text{for no } M' \in A, M'RM\}.$$

This operator μ satisfies the following "propositional logic," as is easily checked:

$$\mu p \wedge \mu q \rightarrow \mu(p \vee q)$$
$$\mu(p \vee q) \rightarrow \mu p \vee \mu q$$
$$\mu p \rightarrow p.$$

(As an exercise, try to derive the further laws $p \wedge \mu q \rightarrow \mu(p \wedge q)$ and $\mu p \rightarrow \mu\mu p$.) Upon due reflection, it may be seen that 1) this logic is complete for its intended interpretation, and 2) it coincides with the modal theory inside the minimal modal logic K (see Chap. 2) of the following operator:

$$p \wedge \Box \neg p.$$

Dynamic Logic

There are also examples of new branches of Intensional Logic which have arisen in computer science. One example is Dynamic Logic, as developed by Vaughan Pratt and others. (For surveys with a logical slant, see R. Goldblatt, 1982, *Axiomatizing the Logic of Computer Programming*, Springer, Berlin & New York (Lecture Notes in Computer Science 130), or also R. Goldblatt, 1986, *Logics of Time and Computation*, CSLI Lecture Notes No. 7, Stanford.)

Dynamic Logic is a historical successor to earlier forms of *operational semantics* for programs, where the latter denote sets of admissible transitions between states of a computer. For instance, an assignment $x := t$ denotes all transitions from one state to another whose new value at the address "x" is the old value of t. In fact, states might be identified with assignments; as happens in the text book D. Gries, 1981, *The Science of Programming*, Springer, Berlin; which is wholly set up in this style, due to Tony Hoare (and widely advocated by Edsger Dijkstra). Behavior of programs can then be described using so-called *preconditions* and *postconditions*, as in the well-known correctness assertion

$$\{\varphi\}\pi\{\psi\} :$$

for each state s_1 satisfying φ, and each successful π-transition (s_1, s_2), φ holds at s_2. A good deal of logical theory has arisen concerning the model theory and proof theory of correctness assertions.

The basic observation in Dynamic Logic (due to Bob Moore) is that there is a modal possible worlds structure in the background. Its universe consists of states, patterned by a *family* of accessibility relations R_π, one for each program π. Then, modalities $[\pi]$ occur, with the obvious meaning:

$$M \models [\pi]\varphi[s] \quad \text{if} \quad M \models \varphi[s'] \text{ for all } s' \text{ with } R_\pi ss'.$$

And the above correctness assertion becomes one of many notions expressible in this formalism:

$$\varphi \rightarrow [\pi]\psi.$$

Incidentally, the logic of such correctness assertions also shows their analogy with the earlier *conditionals*. But, we shall pursue the more general modal perspective here.

The interesting new feature of Dynamic Logic is the interaction of two algebras. One is the usual linguistic structure of Boolean operators with modalities. The other is some selected algebra of programming constructions building complex programs out of components. For instance, we can think of Sequencing, IF THEN ELSE or WHILE DO constructions. For the sake of elegance, these may be replaced by the following:

$$\pi_1; \pi_2 \qquad (sequencing),$$
$$\pi_1 \cup \pi_2 \qquad (\text{indeterministic } choice),$$
$$\pi^* \qquad (\text{Kleene } iteration).$$

In the semantics, we shall require

$$R_{\pi_1;\pi_2} \quad = R_{\pi_1} \circ R_{\pi_2} \qquad (\text{composition of relations}),$$
$$R_{\pi_1 \cup \pi_2} \quad = R_{\pi_1} \cup R_{\pi_2},$$
$$R_{\pi^*} \quad = \bigcup_n R_{\pi^n},$$

starting from some basic assignments to atomic programs.

The modal operator [ ] may be viewed as taking programs and assertions to assertions. Conversely, a *test* operator ? takes assertions φ to test programs φ?, whose semantics is simply

$$R_{\varphi?} = \{(s, s\} \mid M \models \varphi[s]\}.$$

As a result of all these stipulations, program relations can have zero, one or more successor values at different states. This reflects computational practice.

Here are the two promised definitions:

$$R_{\text{IF } \in \text{ THEN } \pi_1 \text{ ELSE } \pi_2} = R_{(\in?;\pi_1)\cup(\neg\in?;\pi_2)},$$
$$R_{\text{WHILE } \in \text{ DO } \pi} = R_{(\in?;\pi)^*;\neg\in?}.$$

These are extensional equivalences, in terms of sets of successful transitions. "Intensional" differences in preferred ways of thinking about the execution of the left-hand and right-hand programs may remain. (These are only captured in more sensitive *process algebras*; cf. J. Bergstra and J. Klop, 1984, Process Algebra for Synchronous Communication, *Information and Control* 60(1/3), 109–137.)

Now, the above semantics for the propositional case generates a basic logic, consisting of 1) basic propositional and modal principles and 2) decomposition principles for complex programs. Evidently, the greatest interest here attaches to the latter kind, such as

$$[\pi_1; \pi_2]\varphi \quad \leftrightarrow [\pi_1][\pi_2]\varphi,$$
$$[\pi_1 \cup \pi_2]\varphi \leftrightarrow [\pi_1]\varphi \wedge [\pi_2]\varphi,$$
$$[\in?]\varphi \quad \leftrightarrow (\in \rightarrow \varphi).$$

For the infinitary notion of iteration, one has certain *induction axioms:*

$$[\pi^*]\varphi \leftrightarrow \varphi \wedge [\pi][\pi^*]\varphi,$$
$$\varphi \wedge [\pi^*](\varphi \rightarrow [\pi]\varphi) \rightarrow [\pi^*]\varphi.$$

Together, these form the so-called *Segerberg Axioms* for Dynamic Logic, whose completeness was proved by various authors (including Rohit Parikh and Dexter Kozen).

It must be observed that completeness fails in certain natural wider settings. For instance, most reasonable formulations of *quantified* Dynamic Logic are highly complex, because True Arithmetic can be coded into them. (The trick is to exploit the *standard* nature of some sequence of states $sR_a sR_a s \ldots$, in defining a copy of the standard natural numbers in their associated individual domains. All that is required can be expressed using the Kleene star. Or more succinctly: as long as iteration is standard, we can describe infinite standard structures in the quantified Logic—with usually disastrous effects on complexity.)

There are other ways of strengthening expressive power, however, which need not endanger completeness or decidability. For instance, George Gargov and his colleagues at Sofia University have observed that one may also view the above as a fragment of a full *modal logic of relational algebra*. A full relational algebra will at least include all *Boolean* operations, such as complement and conjunction, as well as, say, *converse* and perhaps other natural operations. Note that, e.g., conjunction of programs has a plausible interpretation as being one way of enforcing parallel execution. Adding the latter is not a dangerous move: see R. Danecki, 1985, Nondeterministic Propositional Dynamic Logic with Intersection is Decidable, *Lecture Notes in Computer Science* 208, Springer, Berlin, 34–53. But, adding complement too will make the logic very complex, as was shown by the Sofia group.

But semantic questions concerning this system can also be of a newer variety. For instance, one basic distinction is that between *determinism* and *non-determinism* in execution of programs. The usual approach here is to define some class of typically "deterministic" programs, starting from atomic programs denoting total functions, and then using the earlier constructions of sequencing, conditional choice (IF THEN ELSE) and guarded iteration (WHILE DO). It has been shown, for instance, that this system is strictly less expressive than full propositional dynamic logic. (See D. Harel, 1984, Dynamic Logic, in D. Gabbay and F. Guenthner, eds., *Handbook of Philosophical Logic*, vol. II, Reidel, Dordrecht and Boston, 497–604. The proof is a typical example of the combinatorial power needed to extend ordinary modal types of argument to this area.) In addition to this "linguistic" notion of determinism, however, there is also a purely "structural" one. Call any program described in propositional dynamic logic *structurally deterministic* if its denotation is a function in all models whose atomic program relations are functions. All the above programs are structurally deterministic. But is the formalism expressive enough, in that a *converse* holds as well? Other interesting semantic questions concerning dynamic logic abound. For instance, most general model-theoretic concerns have their counterparts in this area. One illustration is the phenomenon of *monotonicity*. A statement $\varphi(p)$ can be monotone in the propositional variable p, in the sense that enlarging the truth range $V(p)$ in a model where $\varphi(p)$ holds at some world, will leave φ true at that world. A statement can also be monotone, however, with respect to some *program* variable a, in the sense that enlarging R_a will preserve its truth. As usual, obvious sufficient conditions can be found for monotonicity, in terms of "positive occurrence" for p and a. (Because of the mutual recursion induced by modalities and test, the syntactic clauses are slightly more involved than in the standard case.) One open question is if there is a Preservation Theorem here, making these syntactic conditions also necessary.

To conclude, we mention a new type of question again. Given some precondition φ and postcondition ψ, we may be interested in the existence of some program or action π guaranteeing the outcome ψ:

$$\text{is there a } \pi \text{ such that } \models \varphi \to [\pi]\psi?$$

For instance, φ might describe a room and the properties of our tools, and ψ the desired state of an open wall-safe. In general, this is a rather hopeless task of "automatic program generation." But, for the restricted language of propositional dynamic logic, there are better results. Notably, the question is *decidable* for the case of programs with only non-modal tests and without iterations. (The search space for π may be bounded, in a suitable manner, by the operator depth of φ. See J. van Benthem, 1987, *Semantics of Programming Languages*, Faculteit Wiskunde en Informatica, University of Amsterdam.) Without these restrictions, the situation becomes more complex. In fact, arbitrary tests trivialize the issue, since always $\varphi \models [\psi?]\psi$. Thus, we have to add a requirement of *termination* then:

$$\varphi \models [\pi]\psi \wedge \langle \pi \rangle \psi,$$

or equivalently,

$$\varphi \models [\pi]\psi \wedge \langle \pi \rangle T.$$

In the earlier simple case, this question still remains decidable. For the more complex cases, nothing definite seems to be known.

Dynamic Logic also has a wider potential, as a general theory of structured *actions*. For an elaboration of this idea in a legal setting, see J.-J. Meyer, 1984, Deontic Logic Viewed as a Variant of Dynamic Logic, report IR-93, Department of Mathematics and Computer Science, Free University, Amsterdam. Then, there are also applications to general dynamic strategies of interpretation for natural language, witness J. Groenendijk and M. Stokhof, 1987, Dynamic Predicate Logic, Department of Philosophy, University of Amsterdam. (See also J. Barwise, 1987, Noun Phrases, Generalized Quantifiers and Anaphora, in P. Gärdenfors, ed., *Generalized Quantifiers: Linguistic and Logical Approaches*, Reidel, Dordrecht and Boston, 1–29.)

Finally, again there is no escape eventually from the *interaction* between dynamic notions and other intensional phenomena, such as knowledge. (See the work by Bob Moore mentioned in Chap. 4.) For instance, Moore considers such principles as

$$[\pi]K\varphi \rightarrow K[\pi]\varphi$$

("if doing π makes me know that φ, then I know that doing π will result in φ").

Semantically, these will reflect certain connections between dynamic and epistemic accessibility relations on states of the world.

Varia

The preceding survey has by no means exhausted the subject of applied Intensional Logic in Computer Science. For instance, certain *general ideas*, rather than specific areas, may be useful too. Thus, the general study of equivalences between intensional axioms and first-order frame conditions may be brought to bear upon recent research into first-order reducible (and hence, computationally more tractable) cases of Circumscription. (See V. Lifschitz, 1986, Computing Circumscription, Department of Computer Science, Stanford University; and J. van Benthem, 1987, Parallels in the Semantics of Natural Languages and Programming Languages, to appear in M. Garrido, ed., 1988, *Logic Colloquium. Granada, 1987*, North-Holland, Amsterdam.)

In fact, the general ideas of possible worlds modelling have a noticeable resilience; witness the recent publication S. Rosenschein and L. Kaelbling, 1987, The Synthesis of Digital Machines with Provable Epistemic Properties, Technical Note 412, SRI International, Menlo Park. The aim of this project is to analyse so-called *situated automata* interacting with their environment. One basic idea here comes from Situation Semantics: the knowledge which such automata can be said to have resides largely in successful attunement to their environment, not in complex internal representation. But, that external localization of knowledge is modelled rather well in Hintikka-type possible worlds semantics with accessibility relations. Hence, it comes as no surprise to find a relatively standard combined epistemic-dynamic logic behind situated automata.

Another area of computer science where modal techniques have been applied is the theory of *data bases*. For instance, various interesting epistemic relations may be defined among individuals, in terms of what information a certain data base contains about them. (See E. Orlowska, 1985, Logic of Indiscernibility Relations, *Springer Lecture Notes in Computer Science* 208, 177–186.) Modal logics of such structures have been studied in D. Vakarelov, 1987, *S4* and *S5* together—*S4+5*, Sector of Mathematical Logic, University of Sofia.

Finally, only very recently, modal structures have emerged in Computational Linguistics. There has been an intensive search for a simple logic of the so-called *feature structures* used in modern grammatical theories with a computational slant. (Cf. S. Shieber, 1986, *An Introduction to Unification-Based Approaches to Grammar*, CSLI Lecture Notes 4, Center for the Study of Language and Information, Stanford.) This logic should combine reasonable expressive power with a still tractable notion of validity. It was found in the form of a modal calculus on feature structures, witness B. Kasper and W. Rounds, 1987, The Logic of Unification in Grammar, *Linguistics and Philosophy*, to appear. Roughly speaking, feature structures themselves may be viewed as possible worlds models for a simple dynamic logic, which can express basic lexical and grammatical regularities. Moreover, there is a natural relation of homomorphic embedding of one such model in another, which reflects the usual *subsumption* order. Thus, the semantics of this modal formalism also turns out to provide a natural definition of *unification*. (Similar modal ideas are developed in G. Gazdar et multi alii, 1987, *Category Structures*, Report CSLI-87-102, Stanford University) This picture stresses an aspect of Intensional Semantics which was somewhat neglected in the preceding chapters. We can *compare* various possible worlds models standing in certain natural structural relations—and enquire, e.g., into possible *transfer* of truth for formulas of specific types.

In view of this recent cascade of applications, it seems safe to conclude that classical Intensional Logic is alive and well.

III   Recent Developments 2: Partiality

The main theme in this second part is "partiality"; a rather vague, but increasingly popular term. Under this heading, we shall present two developments: the "ontological" move from "total" to "partial" objects in our models, as well as the "linguistic" one from total to partial interpretation of our languages in these models. For instance, in the case of time, durationless moments have been replaced by little chunks of time ("periods"), which may be thought of as pieces of incomplete information about the eventual precise location of a moment. But also, partial valuations may interpret some part, rather than all of the proposition letters at each of the temporal locations (whether points or periods). Eventually, we shall discover that the first move represents a shift in our interpretation of the formal semantics, rather than a change in its machinery. Moreover, in the final analysis, the two moves seem quite closely related—as one major source of ontological structures lies in the linguistic construction of canonical Henkin models.

As before, the topic will be developed in a series of specific themes, the first of which continues Chapter 1.

6 Intervals and Events

Literature

Russell, B. 1926. *Our Knowledge of the External World*. London: Allen & Unwin.

Kamp, H. 1979. Instants, Events and Temporal Discourse. In R. Bäuerle et al., eds., *Semantics From Different Points of View*, Berlin: Springer, 376–417.

Van Benthem, J. 1984. Tense, Logic and Time. *Notre Dame Journal of Formal Logic* 25, 1–16.

From Durationless Points To Extended Periods

The motivation for this ontological move is, as usual, both philosophical and linguistic. Philosophically, there is the existence of two complementary views of time (and space). One is "discrete," in terms of points and their aggregates, the other is more "continuous," in terms of extended periods and their ever finer subdivisions. The former perspective has been predominant in mathematics, while the latter seems closer to intuitive common sense conceptions. It is the *interaction* between these two points of view which turns out enlightening, rather than subordination of one paradigm to another. For instance, "Russell's Program" tried to reconstruct the discrete scientific world picture from continuous common sense conceptions—an enterprise which is being rediscovered with a certain frequency.

A more linguistically oriented question is whether natural language presupposes mathematical points of time: the "instant" of wedding, dying? Opinions diverge here (even in our seminar). In any case, throughout the past decade, linguistic semantic arguments have been put forward for switching from evaluation of sentences at points (as in

Chap. 1) to evaluation at extended periods—taking a more complex notion of "truth" at a period as fundamental. (A point of terminology: the term *interval* will be used henceforth, in accordance with current practice. Notice, however, that its connotation of "point set between boundaries" is not implied in this usage.) For instance, on this view, my writing these notes at a period is the fundamental notion, which may only be indirectly related to what is going on (...) at each point within that period, if there be such. Opponents of the new approach often point out that nobody knows how to reduce truth at intervals to truth at their points (I am writing at this period, but not necessarily at all, or most of its points). But rather than being an objection, this observation is actually a strong argument in favour of interval evaluation. Once you start thinking about these matters, the truly mysterious thing is what it means to "write" at a durationless point in time. (I have to admit that distinguished opponents cheerfully deny this. For instance, Terry Parsons claims that it is completely evident how to "determine" truth of a statement at a point, while this is obscure for an interval.)

Specific arguments which have been proposed in the literature are due to

Bennett and Partee: The *progressive* needs interval evaluation. For, the usual Montague-Scott clause (see Chap. 1) makes the wrong prediction that if I am writing a book now, I have been writing a book—since every open interval around the "now" contains open intervals to the left of "now."

Cresswell: *John polished every boot* means neither "at some past point, John polished all boots simultaneously" nor "all boots have been polished by John, at some past time or another," the only available readings in the earlier semantics. Rather, it means that, over some interval, John polished every boot. Likewise, *Dahlia sang and danced* has a clear interval reading meaning something like: some interval was occupied by Dahlia's singing and dancing (these activities neither being necessarily simultaneous nor disjoint). Notice the meaning shift for *and* in this setting.

Dowty: The elusive difference between *simple past* (*P*) and *present perfect* (*PE*) can now be expressed as follows:

$$M \models P\varphi\,[i] \quad \text{if} \quad \textit{for some } j < i, \ M \models \varphi[j] \qquad \underline{\ \ j\ \ }\ \underline{\ \ i\ \ }\ ,$$

$$M \models PE\varphi\,[i] \quad \text{if} \quad \textit{for some } j \textit{ extending beyond } i \textit{ toward the left}$$
$$\textit{but overlapping } i, \ M \models \varphi[j] \qquad \underline{\ \ j\ \ \underline{\ \ i\ \ }}\ .$$

An extensive survey may be found in D. Dowty, *Word Meaning and Montague Grammar*, Reidel, Dordrecht, 1979, where these ideas are also applied to the *aspects*, natural language's way of implying the duration, termination and becoming of events.

Suitable Semantic Structures

Various kinds of interval structure have been proposed as the basic pattern for the new temporal paradigm. All agree on ("total") *predecence* as a primitive, but different additions are found bringing out the extended nature of the new items. The two main choices seem to be

$$\langle I, <, \sqsubseteq \rangle: \quad \text{add } inclusion,$$
$$\langle I, <, O \rangle: \quad \text{add } overlap.$$

Eventually, a good case can be made for having both as primitives; rather than using the not quite satisfactory "definitions"

$$x \ O \ y \leftrightarrow \exists z (z \sqsubseteq x \wedge z \sqsubseteq y);$$

or, going the other way,

$$x \sqsubseteq y \leftrightarrow \forall z (z \ O \ x \rightarrow z \ O \ y).$$

We could proceed now as in Chapter 1, introducing a formal language, and learning more about reasonable conditions on the above structures by studying specific patterns of inference. Here, another road shall be followed (one also quite open to us in the earlier case). In order to narrow down the semantic structures to the intended "temporal" ones, we formulate some minimal constraints:

$<$ is to be *transitive* and *irreflexive*   (a "strict partial order"),
$\sqsubseteq$ is to be *transitive, reflexive,* and
  anti-symmetric   (a "partial order");

and $<, \sqsubseteq$ are to interact as follows:

$$x \sqsubseteq y < z \rightarrow x < z, \quad x < y \sqsupseteq z \rightarrow x < z \qquad (monotonicity),$$
$$x < y < z \wedge x \sqsubseteq u \wedge z \sqsubseteq u \rightarrow y \sqsubseteq u \qquad (convexity).$$

Convexity expresses the assumption that intervals are uninterrupted. (For events, this assumption has been questioned.) Notice that all these conditions have a *universal* form: postulates with "existential import" do not form part of minimal *logical* constraints. For this, and many other subtleties of deriving minimal postulates, see J. van Benthem, *Minimal Conditions for Interval Models*, manuscript, CSLI, 1984.

In the $<, O$-option, the minimal choice seems to be this:

$$< \text{ is } transitive \text{ and } irreflexive,$$
$$O \text{ is } symmetric \text{ and } reflexive;$$

and $<, O$ interact in

$$x < y \rightarrow \neg x \ O \ y,$$
$$x < y \wedge y \ O \ z \wedge z < u \rightarrow x < u.$$

There is a universal principle not occurring in these sets, as it seems to express a "physical" rather than a "logical" constraint on our conception of Time, viz. *linearity*. Its best-known interval form employs the overlap-formalism:

$$x \, O \, y \, \vee \, x < y \, \vee \, y < x.$$

Linearity is indeed satisfied in two more concrete "standard examples":

INT($\mathbf{Z}$):   all bounded intervals of *integers*,
INT($\mathbf{Q}$):   all bounded open intervals of *rationals*.

The scientific victory of the discrete views of Time already shows in the fact that there are no standard names for continuous structures: we have to refer to their point rivals $\mathbf{Z}$ and $\mathbf{Q}$. The latter reference does give us a fully determinate semantic picture (or rather, we have been educated into believing that it does).

Like the point structures of Chapter 1, temporal interval models also suggest some more *global intuitions* concerning the nature of Time. For instance, we formulated a principle of *Homogeneity*: all points in time are "alike." To the latter "horizontal" principle (valid in INT($\mathbf{Q}$), but not in INT($\mathbf{Z}$)), one may now add a "vertical" principle: the temporal universe is "similar" at all levels of detail. Formally, each interval with its inherited relational substructure for its subintervals is to be isomorphic to the whole universe. This principle of *Reflection* expresses a metaphysical view of universes within universes within ..., whose attraction has been felt by many (modern and classical) students of the subject.

Russell's Project

As in the above two examples, point structures give rise to interval structures in an obvious way. But as it turns out, the interval paradigm is equally generous, in that points may be created as ideal *limits* of maximal decreasing sequences of periods

The point ordering then arises by setting $t_1 < t_2$ if some covering period of t_1 precedes some covering period of t_2. In this way, interval structures may be *represented* as patterns of point-set intervals after all.

Upon closer inspection, there are various ways of carrying out this representation, each with its technical peculiarities. (Actually, Russell himself uses the precedence, overlap-framework; see the appendix to this chapter.) In these notes, the reader might merely wish to acquire familiarity with the idea by applying the above construction to the earlier standard examples INT($\mathbf{Z}$) and INT($\mathbf{Q}$) . In the former case, the integers $\mathbf{Z}$ are recovered—but in the latter, $\mathbf{Q}$ becomes enriched to "almost" the reals.

An aside: There is some interest to the "almost," both technically and philosophically. Technically, what happens is that sometimes maximal decreasing nests of open intervals cannot be of the (perhaps expected) form

$$(((\cdots \quad \bullet \quad \cdots)))$$
$$r$$

focusing upon some real number r. For *rational* numbers r, such nests are not *maximal*: it is still possible to add either all left-hand parts (up to r) or all right-hand parts. Thus the rational numbers become "duplicated." There is an intensional feature here of the Russell construction: different ways of approximating a point need not be identified (here, "left" and "right"). (Jens-Erik Fenstad has remarked that similar phenomena occur with non-standard real numbers in non-standard analysis.)

This feature also bears upon a traditional philosophical puzzle called "the moment of change." The fire has been burning, but now it has died out. There must have been an instant, so the problem goes, at which the transition took place, the moment of change: before it, the fire was still burning, after it, it is out. But, as for this moment itself, there seems to be no better reason to say that the fire burnt than that it did not. Hence, we either say both: which violates the law of Noncontradiction, or we say neither: and violate the law of Excluded Middle.

In our present perspective, we note first that this statement of the problem seems to presuppose a point view of time, and not only that, but even that particular one where two adjacent open intervals must have a common boundary. (This is true for *real* time, but not, e.g., for *rational* time. In technical terms, there seems to be an assumption of Dedekind Continuity.) Does this mean that the problem can be dismissed, once we adopt an interval perspective? Not quite, for it returns in the Russell construction. Assuming densely ordered intervals (in some obvious sense), the construction will create a maximal nest of ever smaller "right wings" of the period of burning, as well as a nest of "left wings" of being out. Thus, two immediately adjacent points are created: one last moment of burning, one of first being out—and we should have to stipulate what to do with these. For instance, if we contract them into one (as has been proposed by Kamp), the above problem reemerges in the following new sense. *If* we want the predicates at interval level ("burning," "being out") to make sense at the point level, how are we to do this consistently? In a more technical tense-logical setting, this problem will return below.

Russell's view of the genesis of "public time" has been reconstructed as follows by S. K. Thomason (*Possible Worlds, Times and Tenure*, department of mathematics, Simon Fraser University, Vancouver, 1979.) Our private experiences form small event structures (here, we shall reduce this to: interval structures), which are eventually pooled into one large public experience. Public time is the result of applying the Russell construction to the latter vast interval structure. In addition, Russell considers "private times," preliminary temporal pictures arising when his construction is applied to private experiences. An obvious question, then, raised by Thomason (though not by Russell) concerns the other conceptual route: how is "private experiences ⇒ private times ⇒ public time" related to "private experiences ⇒ public experiences ⇒ public time"?

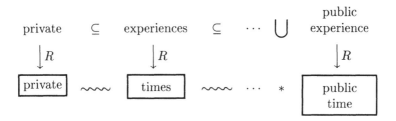

To answer this, the duality is to be studied between interval models standing in certain relations (here: inclusion) and their Russell representations in corresponding relations. For instance, when I_1 grows to I_2, new points may be added, but old points may also be "split up" into several new ones (as new intervals can split formerly bottommost intervals).

When all is said and done mathematically, public time can also be reconstructed as an "inverse limit" of private times. But, interestingly, this particular route will give a *poorer* structure than Russell's original one, lacking a host of instants arising from the global exigencies of public experience in its entirety.

What is becoming clear in these examples is that the main philosophical interest seems to reside in the connections between the various views of time, rather than in some attempted reduction. As usual, by now, this philosophical trend in Intensional Logic also has its linguistic counterpart.

Representation of Temporal Discourse

As part of his larger discourse representation theory, Hans Kamp has developed an account of temporal discourse which combines interval (or event) models with the usual semantic point structures. The basic idea is that a text gives us instructions to build a small discourse model of events with (partial) information about their relations of $<, O$ (and perhaps $\sqsubseteq$). This discourse representation is then related, in a second step, to the traditional point (and point-set interval) structures via "embedding conditions." Thus, the usual picture of

$$\text{TEXT} \underline{\qquad \text{truth definition} \qquad} \text{MODEL}$$

is replaced by the triangle

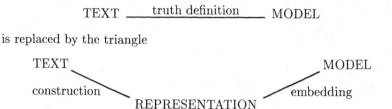

The innovative part here is not the triangle itself (which has been around at least since De Saussure), but the systematic, non-metaphorical implementation: Kamp states construction rules and embedding clauses in exact detail.

As an example of the interplay between interval level and point level, here is Kamp's clever explanation of the difference between the French tenses *imparfait* and *passé simple*. Both of these tenses can describe the very same past event (*Dalida sang*: *Dalida chantait* or *Dalida chanta*), but the speaker has the choice to present the event as "open" or "closed." With an imparfait, it is possible to "break into" the event, adding information about what went on during it—whereas the passé simple presents it as non-extended: "punctually," as the grammatical metaphor goes. But, of course, in terms of denotation, also the passé simple event is extended. This situation is clarified by interpreting the difference as one of instructions for a representation, rather than embedding clauses. Events reported in the passé simple are to function as atomic, non-divisible intervals, whereas the imparfait has no such requirement. (Thus, in the "private time" corresponding to our discourse representation, passé simple events will quite literally be points.) Presumably, the availability of such different "perspectival" instructions plays some useful informative role in discourse—although the why of such matters is still beyond present-day semantic theory.

Partial Model Structures

Texts usually induce only partial information about their event patterns. Accordingly, in the above theory, discourse representations are *partial* interval models

$$\langle I, <^+, <^-, \sqsubseteq^+, \sqsubseteq^- \rangle,$$

with three kinds of information concerning, for example, precedence relationships: $<^+$ (definitely yes), $<^-$ (definitely no), or indeterminate. This "partialization" of classical models is a recurrent folklore idea in philosophical logic. We shall study its logic in Chapter 8 below.

Tense Logic on Interval Models

There have been several attempts to set up a tense logic in the spirit of Chapter 1, but now on interval models—witness L. Humberstone, Interval Semantics for Tense Logics, *Journal of Philosophical Logic* 8, 1979, 171–196; P. Röper, Intervals and Tenses, *Journal of Philosophical Logic* 9, 1980, 451–469; J. Burgess, Axioms for Tense Logics II: Time Periods, *Notre Dame Journal of Formal Logic* 23, 1982, 375–383.

Two main questions arise here, already for the original simple language with $\neg, \vee, \wedge, F, P$:

- What is the meaning of the logical constants in this new setting?

- Which connections exist between truth of a statement at an interval and truth at (all, most, some of) its subintervals?

Both are typical for "partial semantics" in general.

Example 1. Humberstone reads $\neg\varphi$ at an interval as total absence: φ at *no subinterval*; and $\varphi \vee \psi$ as eventual choice: each subinterval has some subinterval with either φ or ψ (but the original interval itself need not make a global choice yet). A useful modal notation here has

$$\square\varphi \text{ is true at } i \quad \text{iff} \quad \varphi \text{ is true } at\ all\ j \sqsubseteq i,$$
$$\Diamond\varphi \text{ is true at } i \quad \text{iff} \quad \varphi \text{ is true } at\ some\ j \sqsubseteq i.$$

Then, the above two readings can also be expressed as $\square\neg\varphi$, $\square\Diamond(\varphi \vee \psi)$ with "ordinary" $\neg, \vee$. Actually, various options seem reasonable here.

Example 2. Humberstone restricts attention to "hereditary" statements φ, preserved in passing from an interval to all its subintervals: $\varphi \rightarrow \square\varphi$. As a general rule, this fails for natural language statements (compare *Dorothea sang "Die Forelle"*)—although it does hold for an important subclass.

Are there any generally valid constraints in this spirit? Humberstone reports a philosophical conviction, expressed by Hamblin (following Prior), that "there are no indefinitely finely intermingled intervals of truth and falsity for any statement." In the above notation, this would seem to say $\neg(\square\Diamond\varphi \wedge \square\Diamond\neg\varphi)$, or equivalently $\square\Diamond\varphi \rightarrow \Diamond\square\varphi$ (i.e., the McKinsey axiom of Chap. 2). There may be something to this, for natural language statements. Röper has a nice observation here: starting from a valuation assigning sets $V(p)$ obeying the above constraint, the *ordinary* readings for $\neg, \wedge, \vee, F, P$ have the effect that the constraint holds automatically for all statements φ.

Actually, both Humberstone and Röper also consider the stronger constraint $\square\Diamond\varphi \rightarrow \varphi$, with the following motivation. If φ fails at i $(\neg\varphi)$, this must be because the interval still harbors the possibility that φ could be excluded $(\Diamond\square\neg\varphi, \text{ or } \neg\square\Diamond\varphi)$. This constraint will no longer hold for the whole language in its traditional interpretation—which leads Röper (and Burgess) to make the move from the "macroscopic" future reading

$$F\varphi \text{ at } i \quad \text{iff} \quad \varphi \text{ at some } j > i$$

to the "micro-future" reading $\Box \Diamond F\varphi$ (F old style). Notice that the latter could be true, even when there is no φ-interval beyond the original one.

Instead of pursuing the logic of these various possible setups, we mention a possible application. First, the variety of choices for truth conditions may reflect a genuine set of options in natural language, and it is important to sort these out. But also, the general constraints may be regarded as describing types of statement. In particular, one is reminded of the "Vendler classification" of verb phrases into *states*, *activities*, *accomplishments* and *achievements*. At least part of the relevant distinctions here can be captured in the above formalism. For instance, states and activities are hereditary, while the other two are not. (But see also below, for a richer perspective.)

Tense Logic at Two Levels

Once a tense logic has been set up at the interval level, the natural question arises as to how it may be related to a tense logic (in the style of Chap. 1) at the level of point representations. Can one find a meaning for the atomic statements at points in such a way that some desirable correspondence results, say:

$$M = \langle I, <, \sqsubseteq, V \rangle \models \varphi[i] \quad \text{iff} \quad R(M) \models \varphi[t] \text{ for all (for most, ...) } t \in i \, ?$$

It is not evident that such an attempt can succeed: perhaps, one eventually wants a different language at the point level. Nevertheless, the issue has been explored, via stipulations such as:

$$p \text{ at } t \quad \text{if} \quad \text{for some } i \in (!) \, t, \, M \models p[i],$$
$$p \text{ at } t \quad \text{if} \quad \text{for all } i \in t, \text{ some } j \sqsubseteq i \text{ has } M \models p[j].$$

On the whole, these speculations have only met with partial success (see *The Logic of Time*, Chapter II.4).

Further Applications

The dual perspective outlined here can be brought to bear upon many more philosophical and linguistic issues. An example of the former is the discussion of Zeno's Paradoxes in *The Logic of Time*, an example of the latter is again the Vendler classification. Achievements (*Dahlia died*) have an instantaneous aspect, even though the event reported can take time: again, a separation between interval representation level (with the event *reported* as atomic) and eventual denotation may provide a clue to the solution.

An extension of the enterprise to *events*, rather than intervals, has also been mentioned at various places. As events have temporal, spatial and "modal" structure, our study will then encompass the topics of Chapter 4 as well. In particular, this point of view has become prominent in Computer Science, witness our brief discussion in Chapter 5, referring to motivations from the areas of distributed computation and artificial intelligence. In the latter area, the case of time is only one example of a larger program, trying to recapture a "Naive Physics" underlying common sense reasoning. (See P. Hayes, 1979, The Naive Physics Manifesto, in D. Mitchie, ed., *Expert Systems*, Edinburgh University Press/ J. Hobbs, ed., 1985, *Commonsense Summer: Final Report*, Report CSLI–85–35, Center for the Study of Language and Information, Stanford University.)

Appendix: The Russell Construction in a NutShell

The following presentation is an elegant reconstruction of Russell's views, due to S. K. Thomason.

a. Axioms: i. $\forall x \ \neg x < x$,

ii. $\forall xyst: x < y \wedge s < t \rightarrow x < t \vee s < y$.

Definition: $x \, O \, y = \neg x < y \wedge \neg y < x$ ("overlap").

b. Example: derive all $<, O$-postulates mentioned before.

c. Let $\langle E, < \rangle$ satisfy the postulates.
Define $\langle T, < \rangle$ as the set of all maximal sets of mutually overlapping objects in E, with $t_1 < t_2$ if $\exists \, e_1 \in t_1, e_2 \in t_2$: $e_1 < e_2$ (in $\langle E, < \rangle$).

d. Claim: $\langle T, < \rangle$ is a strict linear order.
Example: if $t_1 < t_2 < t_3$, then there are $e_1 \in t_1$, $e_2, e_2' \in t_2, e_3 \in t_3$ such that $e_1 < e_2, e_2' < e_3$, and $e_2 \, O \, e_2'$. But then $e_1 < e_3$: and $t_1 < t_3$. (Otherwise: $e_2' < e_2$, by axiom (ii), contradicting $e_2' \, O \, e_2$.)

e. The function $\theta(e) \mapsto \{ t \in T \mid e \in t \}$ ("duration") assigns non-empty convex sets of points to events e in E.
Proof: if $t_1 < t_2 < t_3$, with $t_1, t_3 \in \theta(e)$, then the above situation of e_1, e_2, e_2', e_3 repeats itself, with $e \, O \, e_1, e \, O \, e_3$. We show that e must overlap any event e_4 overlapping both e_2, e_2': whence it must be in t_2 (by maximality). Suppose $e < e_4$ (the other case is similar). We have $e_2' < e_3$, and hence either $e < e_3$ (contradicting $e \, O \, e_3$) or $e_2' < e_4$ (contradicting $e_2' \, O \, e_4$).

f. Moreover, θ is an $<$-isomorphism between $\langle E, < \rangle$ and the interval structures $\langle \{ \, \theta(e) \mid e \in E \, \}, < \rangle$ with $X < Y \Leftrightarrow_{\mathrm{def}} \forall \, t_1 \in X, \, t_2 \in Y: t_1 < t_2$.

g. If the linearity assumption is dropped, the earlier $<, O$-axiom set may be used (with a modified representation!), together with a suitable convexity axiom.

h. Open question: axiomatize the universal $<, O$-theory of non-empty convex intervals on strict partial orders.

i. Similar methods of representation can be used to characterize other proposed kinds of event structure. For instance, Lamport's system works with what may be described as precedence $<$, overlap O and one additional primitive, with an intuitive interpretation

$$X \ll Y \qquad\qquad \exists t_1 \in X, t_2 \in Y : t_1 < t_2.$$

The complete first-order theory of these notions on Lamport's partially ordered event structures is given by the earlier postulates for $<, O$ (without linearity), together with

$$x < y \to \neg y \ll x,$$
$$x \ll y < z \ll u \to x \ll u,$$
$$x < y \ll z < u \to x < u,$$
$$x \, O \, y < z \ll u \to x \ll u,$$
$$x \ll y < z \, O \, u \to x \ll u,$$
$$x \, O \, y < z \, O \, u \to x \ll u.$$

This, then, is the basic background for arguing about concurrent processes. (Usually, such a list will fall out directly from the relevant representation argument.)

j. Thomason has also cast Russell's general views about finite approximations of public time in a categorial framework, with a *category* of event structures related by suitable *homomorphisms*. Interestingly, Lamport has an essentially similar setup for getting coarser and finer "views" of some distributed process.

7 Possibilities and Information

The move toward partial information is also discernible in modal logic; indeed, it has been around there since the sixties. Nevertheless, no coherent set of issues has developed around this theme—and we shall merely present some relevant examples. For a very readable debate on some of the main issues here, see J. Perry, 1986, From Worlds to Situations, *Journal of Philosophical Logic* 15:1, 83–107, and R. Stalnaker, 1986, Possible Worlds and Situations, *Journal of Philosophical Logic* 15:1, 109–123. (These authors stress even one more source of "partiality," in that *any* language we choose, whether fully interpreted or not, is bound to leave out certain aspects of *Reality*.) Some further logical analysis of main themes is attempted in Chapter 8.

Intuitionistic Logic

There is a branch of possible worlds semantics which was interpreted "partially" from its inception, viz. Kripke's analysis of intuitionistic logic. As is well-known, intuitionistic truth seems to have the force of verifiability, or provability. This impression was validated, in a way, by Gödel's early translation of intuitionistic logic into the modal logic *S4*, with $\Box$ read as provable (compare Chaps. 2, 5). Thus, the existing possible worlds semantics for *S4* could be made to serve intuitionistic logic as well, with truth conditions prescribed by Gödel's translation clauses for the logical operators.

Models $M = \langle I, \sqsubseteq, V \rangle$ consist of a partial order $\langle I, \sqsubseteq \rangle$ ("stages of knowledge" with a pattern of possible growth), with a valuation V assigning $\sqsubseteq$-closed subsets of I to proposition letters: if $i \in V(p)$ and $i \sqsubseteq j$, then $j \in V(p)$ ("knowledge is cumulative"). The truth definition now becomes:

$$M \models p[i] \qquad \text{iff} \quad i \in V(p),$$
$$M \models \varphi \wedge \psi [i] \quad \text{iff} \quad M \models \varphi[i] \text{ and } M \models \psi[i],$$
$$M \models \varphi \vee \psi [i] \quad \text{iff} \quad M \models \varphi[i] \text{ or } M \models \psi[i],$$
$$M \models \neg\varphi [i] \qquad \text{iff} \quad \text{for no } j \sqsupseteq i, \ M \models \varphi[j],$$
$$M \models \varphi \rightarrow \psi [i] \quad \text{iff} \quad \text{for all } j \sqsupseteq i, \ M \models \varphi[j] \text{ only if } M \models \psi[j].$$

For instance, the idea of the last clause is that $\varphi \rightarrow \psi$ is true on my present information if, whenever my information grows so as to include φ, it will also include ψ. (In terms of a discussion: being committed to an implication amounts to a promise of defending the consequent as soon as one has incurred the obligation of defending the antecedent.) As in Chapter 6, the meaning of earlier logical constants has to be rethought in such a partial setting: and there are pitfalls. For example, the above clause for $\varphi \vee \psi$ looks "classical," but actually, it is very strong now: one could only be committed to a disjunction (at a given stage of knowledge) by being committed to some disjunct. The "classical" idea here would be rather to promise some choice eventually, perhaps after further relevant facts have been ascertained—moving closer to the reading $\Box\Diamond(\varphi \vee \psi)$ mentioned in Chapter 6.

Notice the ease with which the formalism of Chapter 2 can be reinterpreted "partially." That the latter picture is imperative for intuitionistic logic follows from the intuitionist conception of the mathematical activity as a stepwise creation of knowledge, as well as of the very objects this knowledge is about. This feature would come out more vividly if we were to introduce intuitionistic *predicate* logic, where domains grow along with $\sqsubseteq$ (and even objects may "develop," in some versions of the semantics). In this chapter, however, we shall confine attention to what may be learnt at the propositional level.

The above semantics shows striking similarities with the notion of "forcing" in set theory, as well as certain notions of truth in category theory. The former conglomerate of ideas was developed in detail in M. Fitting, *Intuitionistic Logic, Model Theory and Forcing*, North-Holland, Amsterdam, 1969.

Truth-Value Gaps and Supervaluations

The preceding semantics had a "partial" motivation for its structures, but no "linguistic" partiality, in the sense mentioned at the beginning of Chapter 6. But, the latter direction has also been followed, both in classical and modal settings. For instance, propositional valuations V may be considered with "truth-value gaps," leaving certain propositions without a truth value. (Technically, one moves from total to partial functions from proposition letters to {*true, false*}.) There is a modal flavor here, in that this picture naturally invites us to think of various ways in which a partial V could grow into a total one.

The "supervaluation" idea (due to Van Fraassen in the sixties) is to call those statements true at a partial V which will hold (in the ordinary sense) in all its total extensions. Thus, if V is defined on p, but not on q, statements about p alone will be evaluated as usual—whereas, for example, for q we know only classical *tautologies*, such as $q \lor \neg q$, without being able to infer either q or $\neg q$. (In other words, truth no longer decomposes recursively.) There is an interesting alternative here for the logic of partial reasoning: are we to conclude only what we can see right now, or can we just as well add those conclusions that would be valid, no matter how our ignorance is going to be resolved?

A well-known application of the above ideas is found in K. Fine, Vagueness, Truth and Logic, *Synthese* 30, 1975, 265–300. The "Paradox of the Heap" is the problem, dating back to Antiquity, that there could be no heaps, given that zero grains of sand do not make one, and adding just one grain of sand to a non-heap will not turn it into one. Mathematical induction then seems to preclude the existence of heaps of sand at all, with any number of grains. Fine takes this to be a problem of vagueness: the predicate "heap" definitely applies to large numbers of grains (say, over a billion), and definitely does not apply to small numbers (say, under a thousand). In between is a grey zone, with truth value gaps, which can only be made black-and-white by stipulation. But then, with the above supervaluation account of truth, we shall have to call the second premise *false*: on any total valuation extending our partial one, there will be a "transition number" which is not a heap, although the next higher one is. (Of course, this number is not the same in all such valuations—but that is not required.)

Actually, Fine's solution is just one among many. The Paradox of the Heap continues to inspire new accounts of vagueness, more recent ones also taking into account the *context-dependent* character of predicates such as "heap." (What one would call a heap depends on the size of the objects in a "reference environment.") But, the latter theme lies beyond the scope of these notes.

Already, these two examples from an early era in philosophical logic induce several questions about partiality, and the relationships between its various manifestations. For instance,

- What is the connection between the more "structural" and more "linguistic" approaches to partiality?

In particular also, is there a possible trade-off between truth-value gaps and failure of Excluded Middle (as in the intuitionistic case) while still interpreting the whole language?

- What reductions exist between truth at partial locations and truth at their total completions?

In particular again, how can we use the latter to "speed up" reasoning at partial locations?

All these themes will be considered in Chapter 8 below. Here, we turn to a short survey of some more recent pleas for partiality within the framework of possible world semantics.

From Worlds to Possibilities

The change in perspective advocated for tense logic in Chapter 6 has been extended to modal logic in L. Humberstone, From Worlds to Possibilities, *Journal of Philosophical Logic* 10, 1981, 313–344. "Possibilities" are fragments of total possible worlds, say "states of affairs" or "local histories." The technical elaboration of these views employs models $\langle W, R, \sqsubseteq, V \rangle$, where possibilities are now ordered both by alternativity and inclusion. In a sense, different extensions of some possibility may be regarded as modal alternatives too; but this is not enough: *conflicting* possibilities may have to be admitted as well. No new logical issues emerge in Humberstone's paper—but most of the earlier ones concerning tense also turn out to make good sense in a modal perspective.

One new type of argument for this approach in the modal case should be mentioned. In traditional treatments of knowledge and belief (following Hintikka), my believing (knowing) that φ at w always involves some *set of* doxastic (epistemic) alternatives. Now this multiplicity often seems forced upon us: there is just me, my beliefs (and the world I am in)—and the "set of" arises only because so many total worlds may be compatible with those (partial) beliefs. Humberstone suggests that, in these cases, *one* possibility may do the job of many possible worlds. (This idea had already been proposed in the early seventies by Fine, for the semantics of relevant logics.) Although this may be plausible in some cases, it should be admitted that the "variation" assumed in Hintikka's semantics also has certain intuitive attractions in describing the uncertainties of actual belief.

Data Semantics

A more inventive approach in the above spirit, which tries to get maximal modal leeway out of the $\sqsubseteq$-patterns only, is the "data semantics" of F. Veltman, Data Semantics, in J. Groenendijk et al., eds., *Formal Methods in the Study of Language*, Mathematical Centre, Amsterdam, 1981. (Reprinted in the GRASS-series, Foris, Dordrecht, 1984, Vol. II.)

Veltman starts from an ontology of "facts" forming "data sets," which can *confirm* or *refute* statements. In its most accessible formulation, the resulting semantics looks as follows. Models are triples $\langle W, \sqsubseteq, V \rangle$, where $\langle W, \sqsubseteq \rangle$ is a *partial order* ("information stages"), subject to the additional constraint that every maximal chain have a greatest element ("optimal searches are rewarded"). (The latter condition is reminiscent of the CPO's in Dana Scott's domain semantics.) The valuation function V

assigns partial propositional valuations V_w to each $w \in W$, in such a way that V_u extends V_w when $w \sqsubseteq u$ ("monotonicity") and V_w is total for topmost elements w in the order. Thus, data models contain their own completions, in a sense—a reasonable, but not an exclusive option.

The truth definition now extends the above behavior, with an interesting presentation in terms of positive and negative turnstiles:

$$\begin{cases} M \models^+ p[w] & \text{if} \quad V_w(p) = 1 \\ M \models^- p[w] & \text{if} \quad V_w(p) = 0, \end{cases}$$

$$\begin{cases} M \models^+ \varphi \wedge \psi\, [w] & \text{if} \quad M \models^+ \varphi[w] \ and \ M \models^+ \psi[w] \\ M \models^- \varphi \wedge \psi\, [w] & \text{if} \quad M \models^- \varphi[w] \ or \ M \models^- \psi[w], \end{cases}$$

$$\begin{cases} M \models^+ \varphi \vee \psi\, [w] & \text{if} \quad M \models^+ \varphi[w] \ or \ M \models^+ \psi[w] \\ M \models^- \varphi \vee \psi\, [w] & \text{if} \quad M \models^- \varphi[w] \ and \ M \models^- \psi[w]. \end{cases}$$

The interesting difference with the earlier intuitionistic case is in the behaviour of negation. It merely flip-flops:

$$\begin{cases} M \models^+ \neg\varphi\, [w] & \text{if} \quad M \models^- \varphi[w] \\ M \models^- \neg\varphi\, [w] & \text{if} \quad M \models^+ \varphi[w]. \end{cases}$$

(As many people have pointed out, this presentation, and even these very clauses—apparently first used by Hans Kamp—is equivalent to a traditional *three-valued* scheme with values 0, 1, "Undefined." Still, it has some heuristic virtues.)

Now, Veltman goes on to consider modal notions, exploiting the branching pattern of valuations. His aim here is to provide formal equivalents for natural language *if then*, *must* and *may* (rather than some philosophical relatives thereof).

$$\begin{cases} M \models^+ \varphi \Rightarrow \psi\, [w] & \text{if} \quad for\ all\ u \sqsupseteq w: \\ & \qquad M \models^+ \varphi[u] \ only\ if\ M \models^+ \varphi[u] \\ M \models^- \varphi \Rightarrow \psi\, [w] & \text{if} \quad for\ some\ u \sqsupseteq w: \\ & \qquad M \models^+ \varphi[u] \ and\ M \models^- \psi[u], \end{cases}$$

$$\begin{cases} M \models^+ \text{MUST}\, \varphi\, [w] & \text{if} \quad M \models^+ \varphi[u] \ at\ all\ endpoints\ u \sqsupseteq w \\ M \models^- \text{MUST}\, \varphi\, [w] & \text{if} \quad M \models^- \varphi[u] \ for\ some\ endpoint\ u \sqsupseteq w, \end{cases}$$

$$\begin{cases} M \models^+ \text{MAY}\, \varphi\, [w] & \text{if} \quad M \models^+ \varphi[u] \ for\ some\ u \sqsupseteq w \\ M \models^- \text{MAY}\, \varphi\, [w] & \text{if} \quad M \models^- \varphi[u] \ for\ all\ u \sqsupseteq w. \end{cases}$$

(Various readers have pointed out that this is very close in spirit to an *intuitionistic S4* logic, especially since the points of divergence, e.g., in the $\models^-$-clause for $\Rightarrow$, are debatable. Nevertheless, again, this particular presentation has led to applications apparently hidden in the former framework.)

The resulting "data logic" is axiomatized in Veltman's dissertation (cf. Chap. 3). One interesting feature is the modal logic of MUST and

MAY, which retains those properties of modal $\square, \lozenge$ which seem valid for their ordinary language counterparts (such as $\square(\varphi \wedge \psi) \leftrightarrow \square\varphi \wedge \square\psi$, $\square\varphi \leftrightarrow \neg\lozenge\neg\varphi$, $\square\varphi \rightarrow \square\square\varphi$), while also accounting for a crucial difference. Recall the discussion in Chapter 2: MUST φ may be weaker, or rather, less "direct" than φ. This difference shows up in the above semantics: when MUST φ is true, all avenues of investigation end up with φ—but, this is not (yet) sufficient to enforce truth of φ itself. In other words, Veltman denies the Humberstone-Röper condition $\square\lozenge\varphi \rightarrow \varphi$ (Chap. 6).

Another feature explored systematically by Veltman is *monotonicity*. Statements can be "+ stable," when their truth, once established, remains unchanged: examples are p, $p \Rightarrow q$ and a non-example is MAY p. Statements can also be "− stable": once false they remain that way. Examples are: p, MAY p, a non-example is $p \Rightarrow q$. And there are statements which are neither, such as MAY $p \Rightarrow$ MAY q. These phenomena can be studied and classified, with rather interesting questions arising (such as: how often can a statement "change its mind"?).

Partial Views of Conditionals

A similar move has been proposed in the semantics for (counterfactual) conditionals, in R. Turner, Counterfactuals without Possible Worlds, *Journal of Philosophical Logic* 10, 1981, 453–493. Briefly, Turner's motivation is that evaluating conditionals does not involve comparing whole worlds (or world courses), as is suggested in the Lewis semantics of Chapter 3, but rather small local descriptions of situations retaining some relevant features of our world. These he orders by a relation of "relative plausibility," whence the basic idea of Lewis remains operative. (There are many intricacies in Turner's setup, some of them of possible independent interest. For instance, one suggestion emanating from his approach is that we take growing "possibilities" to be ever more detailed finite descriptions of alternative possible worlds, with the steps measured by increasing *quantifier depth*, as in Hintikka's normal forms for predicate logic.)

As in the modal case with alternativity and inclusion, it is of some interest how far one can go without using relative plausibility, exploiting the $\sqsubseteq$ pattern only. For instance, in data semantics, one could define a counterfactual implication by having $\varphi \Rightarrow^* \psi$ true at w if, at the largest $u \sqsubseteq (!)w$ which still has extensions validating φ, $\varphi \Rightarrow \psi$ holds in the original sense. Intuitively, this backing-up procedure seems to presuppose access to a unique "epistemic past," either deictically, or as a consequence of a tree-like data structure to begin with. (This particular proposal for evaluating conditionals was already made by Ray Jennings in 1981 at the Halifax meeting of the Society for Exact Philosophy.)

In the last analysis, these partial analyses move closer towards the original *syntactically* oriented approaches of pre-Kripkean times (as is particularly evident in Turner's work). The connection between "partial semantics" and "ordinary syntax" will be one of the themes in the next chapter.

Further Proposals

Partiality has already become something of an article of faith in some recent theories, such as the *Situation Semantics* of J. Barwise and J. Perry, *Situations and Attitudes*, MIT Press, 1983. A detailed comparison of their framework of "situations," "support," "strong consequence," and "constraints" with the developments chronicled above will not be undertaken here. There are obvious analogies, especially in the recent *SS*-exegesis of various logicians, such as H. Kamp, A Scenic Tour Through the Land of Naked Infinitives, *Linguistics and Philosophy*, 1984, to appear; and T. Langholm, Some Tentative Systems Relating to Situation Semantics, in *Report of an Oslo Seminar in Logic and Linguistics*, Mathematical Institute, University of Oslo, preprint series 9, 1984.

All these proposals go beyond the present propositional perspective, in that they employ some versions of partial *predicate* logic. The latter supports partial information about predicates; but eventually one could introduce partiality of *individuals* too. (For instance, the stepwise refinements of events or intervals, as described in the Russell story of Chapter 6, is already one instance of a concrete process where individuals at one stage of knowledge may have several more determined counterparts at the next.) An exploration of such possibilities may be found in F. Landman, 1986, *Towards a Theory of Information. The Status of Partial Objects in Semantics*, Foris, Dordrecht and Cinnaminson, (GRASS series, vol. 6).

Conclusion

This chapter has given us a perhaps bewildering variety of partial semantics. Moreover, motivations for going this way are quite diverse, ranging from more *ontological* to more *epistemic* ones. The latter trend is a very general one in modern semantics: models are becoming thought of as "cognitive structures." Even so, one might expect the same eventual development as described in Part I. The various analogies and cross-connections presented in this chapter may be signs that a condensation could take place to one simple, stable, formal framework, which still leaves open quite diverse ideological uses. It will be interesting to see the logic textbooks of a decade hence.

8   The Logic of Partiality

Certain general questions recurred in Chapter 7 concerning the nature of partiality and its connections with earlier approaches. A number of these will be discussed now, in the light of some related phenomena in traditional logic. Our main conclusions will be that the locus of partiality is diffuse, and that classical methods of handling it take us a long way.

Literature

Van Benthem, J. 1981. Possible Worlds Semantics for Classical Logic. Report ZW-8018, Department of Mathematics, Rijksuniversiteit, Groningen.

Van Benthem, J. 1986. Partiality and Nonmonotonicity in Classical Logic. *Logique et Analyse* 29, 225–247.

"Partial" Models for Classical Logic

In the preceding chapter there seemed to be an interplay between "linguistic" partiality, where not all statements of the language obtain a truth-value (or whatever suitable denotation), and "ontological" partiality, where the indices of evaluation in the models are thought of as partial objects. One reason for the easy switch between these two points of view is the *Henkin construction* in logic, which allows one to transform a syntactic setting of (partial) information sets into model structures. For instance, in the standard Henkin completeness proof for the minimal modal logic K (see Chap. 2), possible worlds are introduced as *maximally* (K-) *consistent sets* of formulas—and in its "possibility" version (see Chap. 7), possibilities would be just consistent sets of formulas. This observation fits in with the earlier remark, made in connection with Turner's partial model analysis of conditionals, that the progress towards *partial semantics* also resembles a return to good old *syntax*.

How much there is to these feelings may be clarified by reconsidering ordinary Henkin models. (The following discussion presupposes some familiarity with this method of proof in ordinary logic.)

There is something inelegant to an ordinary Henkin argument. One has a consistent set of sentences S, perhaps quite small, that one would like to see satisfied semantically. Now, some arbitrary *maximal* extension S^+ of S is to be taken to obtain a model (for S^+, and hence for S)—but the added part $S^+ - S$ plays no role subsequently. We started out with something partial, but the method forced us to be total. (Similar problems surround the use of *ultrafilters* in constructing model-theoretic ultraproducts. For instance, this makes the concept of "the" non-standard reals rather blurred.)

The usual reason given is that only maximally consistent collections behave like semantic truth sets: $\neg\varphi \in S^+$ if and only if $\varphi \notin S^+$, $\varphi \wedge \psi \in S^+$ if and only if $\varphi \in S^+$ *and* $\psi \in S^+$, etc. But, in a suitable partial setting, decomposition occurs just as well. Consider the universe of all consistent sets of formulas, ordered by inclusion. Then we have, for classical derivability $\vdash$,

$$\begin{array}{lll} S \vdash \varphi \wedge \psi & \text{iff} & S \vdash \varphi \text{ and } S \vdash \psi, \\ S \vdash \varphi \to \psi & \text{iff} & \text{for all } S' \supseteq S, \ S' \vdash \varphi \text{ only if } S' \vdash \psi, \\ S \vdash \neg\varphi & \text{iff} & \text{for no } S' \supseteq S, \ S' \vdash \varphi, \end{array}$$

and, for example, $\varphi \vee \psi$ gets the earlier "$\Box\Diamond$"-decomposition. (These observations can be extended to predicate logic: there is no mere accident here.)

As an immediate result, completeness follows for classical logic with respect to a possible worlds semantics having models $M = \langle W, \sqsubseteq, V \rangle$; where $\langle W, \sqsubseteq \rangle$ is a partial order, and the valuation V satisfies

1. $M \models \varphi[w]$, $w \sqsubseteq v$ only if $M \models \varphi[w]$ (Heredity),

2. $M \not\models \varphi[w]$ only if for some $v \sqsupseteq w$, $M \models \neg\varphi[v]$.

Both these principles occurred in Chapter 6 above, in the modal guises $\varphi \to \Box\varphi$, $\neg\varphi \to \Diamond\Box\neg\varphi$ (or $\Box\Diamond\varphi \to \varphi$), respectively. Another way to think about (2) is as follows: if every $v \sqsupseteq w$ has some extension where φ holds, then there is "no escape," and we may just as well have φ in v. This validates the classical principle of Double Negation $\neg\neg\varphi \to \varphi$.

Without (2), but retaining (1), this semantics becomes our earlier *intuitionistic* semantics of Chapter 7. Dropping even (1), that is the presupposition of "monotonicity," we obtain what is perhaps the basic non-monotonic logic, which is axiomatized in the L & A paper mentioned above.

This perspective invites various other relevant questions, of which we mention a few.

First, it is possible to obtain the earlier distinction between "direct" and "indirect" evidence, by distinguishing between the possibilities

"$\varphi \in S$" and "$S \vdash \varphi$" (or, equivalently, for all $S' \supseteq S$ there exists some $S'' \supseteq S'$ with $\varphi \in S''$).

Next, the above construction also works with restrictions on the consistent sets being considered. In particular, only the *finite* ones need be considered, banning maximally consistent sets from sight altogether. The latter could then be *added*, if desired, without disturbing truth and falsity in the original model.

This again raises issues of representation, or at least, the interplay of consistent fragments S and maximally consistent ideal completions S^+. In one sense, there is a very simple relation here (compare the related questions in Chap. 7), through the well-known equivalences

- $S \vdash \varphi$ if and only if $\varphi \in S^+$ for all maximally consistent $S^+ \supseteq S$,

- $\varphi \in S^+$ if and only if $S \vdash \varphi$ for some finite consistent $S \subseteq S^+$.

Reading $\vdash$ as "partial verification" and $\in$ as "truth in," the Completeness Theorem for propositional logic may then even be regarded as a reductive correspondence between truth in a partial setting and truth at sets of total locations.

With another very intuitive picture, however, the situation becomes more tricky. Following the Russell representation (see Chap. 6), one may think of *maximal chains* through the inclusion order as producing total models. This is indeed possible, by letting proposition letters be true if they occurred in the chain, and false otherwise. But, the earlier equivalence, now in the form

$$\text{"total model"} \models \varphi \quad \text{iff} \quad \text{"some chain stage"} \models \varphi,$$

may fail in general. (For instance, the chain $\{p_2\}$, $\{p_2, p_3\}$, $\{p_2, p_3, p_4\}$, ... is easily extended to a maximal chain in the universe of finite consistent sets, none of whose members contains the proposition letter p_1. Thus, although $\neg p_1$ will be true in its chain model, at no particular stage, can p_1 already be excluded.) This reduction will only hold for so-called *generic* branches, making up their mind explicitly about every choice $\varphi, \neg\varphi$.

Finally, the above model invites the introduction of *modal* operators $\Diamond, \Box$, as in earlier accounts of partial information semantics in Chapter 7. The resulting modal logic becomes a kind of intuitionistic *S4*: see the earlier-mentioned L & A paper for some delightful complications.

"Partial" Validity Tests in Classical Logic

Perhaps the most elegant validity test in classical logic is the use of "Beth Tableaus." Here, one systematically searches for a *counterexample* to an inference from $\varphi_1, \ldots, \varphi_n$ to φ by assigning $\varphi_1, \ldots, \varphi_n$ the value 1, φ the value 0, and then decomposing according to the obvious rules. If the procedure leads to conflicting desiderata in all cases, the inference was valid—if not, then the particular branch in the search tree where no conflict occurred will produce a counterexample.

Example (a "closed tableau"):

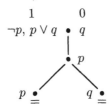

The inference is valid.

Example (an "open branch"):

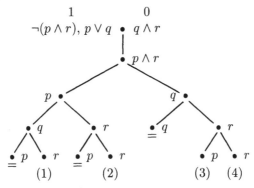

Branches (1), (2), (3) and (4) provide counterexamples.

Such structures, especially in their predicate-logical formulation, have attracted various authors, as a model for actual representation of thoughts. Beth tableaus are, if one wishes, a form of discourse representation (see Chap. 6) with a little inference added. (Compare J. van Benthem and J. van Eyck, 1982, The Dynamics of Interpretation, *Journal of Semantics* 1, 3–20.) Beth even cooperated with the famous psychologist Piaget on these matters.

That closed tableaus imply valid inferences follows by noticing the validity of the following inferences corresponding to bottommost nodes:

$$\varphi_1, \ldots, \varphi_n, \alpha \models \psi_1, \ldots, \psi_n, \alpha$$

(read as: $\varphi_1 \wedge \ldots \wedge \varphi_n \wedge \alpha \models \psi_1 \vee \ldots \vee \psi_n \vee \alpha$), and working upward via the *Gentzen sequent* rules:

$$\neg: \quad \frac{\Phi, \alpha \models \Psi}{\Psi \models \Psi, \neg\alpha} \qquad\qquad \frac{\Phi \models \Psi, \alpha}{\Phi, \neg\alpha \models \Psi}$$

$$\vee: \quad \frac{\Phi, \alpha \models \Psi \quad \Phi, \beta \models \Psi}{\Phi, \alpha \vee \beta \models \Psi} \qquad \frac{\Psi \models \alpha, \beta, \Psi}{\Psi \models \alpha \vee \beta, \Psi}$$

and likewise for conjunction ($\wedge$).

Open branches lead to counterexamples, by the obvious stipulation that $V(p)$ be 1 for those p occurring on the left-hand side, and 0 for those on the other side. This stipulation for proposition letters turns out to extend to all formulas on the branch, by an induction on their complexity. But, these valuations may be *partial*—as was the case with branch (2) in the above example: no value was given to q (it did not matter).

Thus, Beth tableaus employ partial valuations (as in Chap. 7), in the very heartland of logic. How can it be, then, that classical logic results from them, rather than some "strong consequence"? The reason is that, in this setting, at least *two* notions of consequence, still equivalent in the total case but no longer here, need to be distinguished:

> For all valuations V,
> if $V(\text{premises}) = 1$, then $V(\text{conclusion}) = 1$

which is indeed a form of "strong consequence," versus

> For all valuations V,
> if $V(\text{premises}) = 1$, then $V(\text{conclusion}) \neq 0$.

Beth tableaus test the latter notion, producing classical logic.

Nevertheless, the former notion—which (in this limited realm) is equivalent to strong consequence in data semantics (or situation semantics, on Kamp's account)—can also be described quite easily using the tableau idea. Reinterpreting the right-hand side as *not-true*, all Gentzen rules remain valid, except for the above rule

$$\frac{\Phi, \alpha \models \Psi}{\Phi \models \Psi, \neg\alpha}$$

To make up for the latter loss, all classical interchange principles remain valid:

$$\neg\neg\varphi \leftrightarrow \varphi \qquad \text{(Double Negation)},$$
$$\neg(\varphi \vee \psi) \leftrightarrow \neg\varphi \wedge \neg\psi, \quad \neg(\varphi \wedge \psi) \leftrightarrow \neg\varphi \vee \neg\psi \qquad \text{(De Morgan)}.$$

The resulting set of principles is easily seen to completely describe strong consequence. When all is said and done then, it is not all that different from its classical relative. Moreover, this outcome will not change appreciably for the case of predicate logic.

Modal Reductions of Partial Semantics

The various partial semantics reviewed in the preceding chapter had a modal flavor, with the possibility of a trade-off between partial "$\models^+ \varphi$, $\models^- \varphi$, φ undecided" and total (modal) "$\models \Box\varphi$, $\models \Box\neg\varphi$, $\models (\neg\Box\varphi \wedge \neg\Box\neg\varphi)$." As a case in point, we determine the modal logic behind

Veltman's data semantics. (A full proof is in the earlier-mentioned L & A paper.)

First, the modal logic of data structures is an old acquaintance, viz. $S4.1$: $S4$ plus the McKinsey Axiom $\Box\Diamond\varphi \rightarrow \Diamond\Box\varphi$ (see Chap. 2). As an exercise, it is useful to check the *soundness* of these modal axioms. *Completeness* follows by standard, though non-elementary modal methods.

The maximal chain clause for data models is reminiscent of the ω-chain limit clause in Scott's domain semantics. Nevertheless the two are not equivalent: e.g., the uncountable endless well-order $\omega_1, \leq$ is a CPO in Scott's sense, but not an admissible data structure. By the techniques mentioned just now, the modal logic of CPOs remains the old $S4$.

Then, a translation is needed from "data language" to the modal language, reflecting the above observation. A simultaneous recursion is appropriate in this partial setting:

$$
\begin{array}{llll}
(p)^+ & = \Box p & (p)^- & = \Box\neg p, \\
(\varphi \wedge \psi)^+ & = \varphi^+ \wedge \psi^+ & (\varphi \wedge \psi)^- & = \varphi^- \vee \psi^-, \\
(\varphi \vee \psi)^+ & = \varphi^+ \vee \psi^+ & (\varphi \vee \psi)^- & = \varphi^- \wedge \psi^-, \\
(\varphi \Rightarrow \psi)^+ & = \Box(\varphi^+ \rightarrow \psi^+) & (\varphi \Rightarrow \psi)^- & = \Diamond(\varphi^+ \wedge \psi^-), \\
(\text{MUST } \varphi)^+ & = \Box\Diamond(\varphi)^+ & (\text{MUST } \varphi)^- & = \Diamond\Box\varphi^-, \\
(\text{MAY } \varphi)^+ & = \Diamond\varphi^+ & (\text{MAY } \varphi)^- & = \Box\varphi^-.
\end{array}
$$

Now it turns out that data models M can be transformed into $S4.1$-models M^*, by choosing a suitable valuation V^*, such that

$$
\begin{array}{llll}
M \models^+ \varphi[w] & \text{implies} & M^* \models \varphi[w], \\
M \models^- \varphi[w] & \text{implies} & M^* \models \varphi^-[w].
\end{array}
$$

And a converse reduction is feasible as well.

In all then, we have the following:

Theorem: φ strongly follows from premises π in data semantics if and only if $\pi^+ \rightarrow \varphi^+$ is a theorem of $S4.1$.

One technical advantage of such a reduction is that existing modal knowledge about the latter logic can now be transferred to data semantics. For instance, $S4.1$ is *decidable*, and hence so is strong consequence. Thus, traditional modal notions and methods seem to retain their interest, even in a partial environment.

Up till now, our aim was mainly to show how partial logic is still firmly embedded in more traditional logical theories. Nevertheless, we should also be open to various *new questions* suggested by partiality, which might not have come to the surface otherwise.

Advance Reasoning in Partial Settings

Our first example in this spirit has a somewhat special flavor, inspired by the earlier mentioned move from total models $\langle I, < \rangle$ to partial models

$\langle I, <^+, <^- \rangle$ in tense logic. We shall stick with this particular example. On a partial view, the original analysis of total models may quite well be valid as a limiting case. Indeed, it may be investigated how the original class of total models influences the partial model theory. Several issues emerge in this light.

A set of conditions T on models $\langle I, < \rangle$ in the original language determines a class $\mathrm{MOD}(T)$ of admissible models, our ideal realm. This class already fixes a partial realm, viz. those models $\langle I, <^+, <^- \rangle$ which can still "grow into" a model in $\mathrm{MOD}(T)$. (This growth may have two aspects: acquiring *new individuals* as well as *new information* about relations.) Thus, a clear-cut question arises: which set T^* will define the "partial hull" of $\mathrm{MOD}(T)$? The answers will merely be stated here. Take all logical consequences of T which have *universal* form (compare Chap. 6), and apply the following transformation:

- interchange $\neg$ until all negations are in the form of atomic formulas (suppressing $\neg\neg$),

- replace largest subformulas $x < y$ by $\neg x <^- y$, $\neg x < y$ by $\neg x <^+ y$. Call the resulting "partial form" of $T : \pi(T)$.

Theorem: A partial model verifies $\pi(T)$ if and only if it can be extended to some model for T.

A related question is which partial model is already "safe," in that not some, but *all* total extensions (on its domain) are models for T? For *universal* first-order conditions T (as in Chap. 6), there is an answer here: involving a translation π^+ sending $x < y$ to $x <^+ y$ and $\neg x < y$ to $x <^- y$.

But also *conversely*, starting from conditions $S(<^+, <^-)$ in the partial, rather than the total language, $\tau(S) = S(<, \not<)$ may be read as a constraint on the limiting total models. One obvious issue here is whether S-models always admit of extension to total models for $\tau(S)$. Warning example:

$$\forall xy(\neg x <^+ y \land \neg x <^- y) \text{ is consistent, but its } \tau\text{-version is not.}$$

The answer is contained in the above: does S imply $\pi\tau(S)$?

Our final question again concerns reasoning. Suppose that one is in some partial situation (say, a Kamp discourse model), being acquainted with some (universal) postulates T that are going to obtain in whatever total picture arises eventually. How much of an argumentative "advance" can be taken out at this stage already?

Example: Let $I = \{0,1\}$, $<^+ = \{\langle 0,1 \rangle\}$, $<^- = \emptyset$ with $T = \{\forall xy(x < y \rightarrow \neg y < x)\}$. Although $<^-$ is empty, it may concluded now already that $\langle 1,0 \rangle$ could *never* belong to $<^+$, and hence $\langle 1,0 \rangle$ must be in $<^-$. (And likewise for $\langle 0,0 \rangle, \langle 1,1 \rangle$.)

A corresponding notion of consequence, for *quantifier-free* partial language statements φ, ψ and total T, is as follows:

$\varphi \models_T \psi$ if, for all partial models $I_0 \models \varphi$, ψ holds in all total models I extending I_0 which satisfy T. (Here, I interprets $<^+$ as $<$ and $<^-$ as $\not<$, in its role as a partial model.)

Theorem: "Advance Consequence" can be effectively axiomatized.

Proof: Actually, a reduction is possible here to ordinary inference. Let P_φ be the set of those positive formulas α (constructed from atomic formulas $x <^+ y$, $u <^- v$ with $\wedge, \vee$ only) which are valid consequences of φ. Then $\varphi \models_T \psi$ holds if and only if, classically, $\forall xy(x <^+ y \vee x <^- y)$, $\forall xy \neg(x <^+ y \wedge x <^- y)$, T, P_φ imply ψ. ∎

Three-Valued Logic

Behind many of the earlier proposals for partial semantics, there lies a very traditional logical system, viz. a *three-valued* logic. Its truth values are 1 (True), 0 (False) and $*$ (Undefined). Sometimes, even *four* values may be useful, including a value Overdefined (both true and false). Now, "Many-Valued Logic" is a rather stuffy subject, dating back to early decades of this century, which has always led a rather marginal existence. Are we merely on a nostalgic journey?

In fact, there is a remarkable revival of many-valued logics these days, both in mathematical and philosophical areas of logic. See S. Feferman, 1984, Towards Useful Type-Free Theories I, *Journal of Symbolic Logic* 49, 75–111 for foundational aspects, or S. Blamey, 1986, Partial Logic, in D. Gabbay and F. Guenthner, eds., *Handbook of Philosophical Logic III*, Reidel, Dordrecht and Boston, 1–70, for philosophical motivation. Compare also Alasdair Urquhart's survey of Many-Valued Logic in that same volume.

Perhaps the most fundamental logical question in this setting concerns the choice of *primitive connectives*. With three values, there are several natural kinds of operators on propositions which we might want to have available in reasoning. This calls for a number of Functional Completeness results, to be presented below. First, we repeat what are the natural truth tables for "classical" connectives, implicit in preceding sections:

$\neg$			$\wedge$	1	0	$*$		$\vee$	1	0	$*$
1	0		1	1	0	$*$		1	1	1	1
0	1		0	0	0	0		0	1	0	$*$
$*$	$*$		$*$	$*$	0	$*$		$*$	1	$*$	$*$

Note two properties of these schemata:

- they are *conservative*, in that their behavior on classical truth values is as in the standard case;

- they are *persistent*, in that a truth value 0 or 1 once assigned will not change any more when we change arguments from undefined (∗) to defined (0 or 1).

Formulas defined using these connectives will then also be persistent in an obvious sense, as well as *closed*: on total valuations (assigning only 0 or 1), their truth value will be defined (i.e., not ∗).

These are reasonable properties. Nevertheless, it has often been argued that, for instance, natural languages really have *two kinds* of negation. One is a strong denial ("definitely not"), as in the above, the other merely claims absence of truth:

$$
\begin{array}{c|c}
\sim & \\
1 & 0 \\
0 & 1 \\
\ast & 1
\end{array}
$$

Note that a formula like $\sim p$ is no longer persistent here! Thus, we get a system of formulas with diverse logical properties: some are stable under growth of information, others are not. This seems to reflect the realities of life.

With these connectives at hand, some definability results can now be formulated (of which the first and third are folklore).

Proposition: Every closed truth function is definable using $\{\neg, \wedge, \vee, \sim\}$.

Proof: This is an easy matter of recording the truth table, as in the classical two-valued case. ∎

Theorem: (Blamey). Every persistent truth function is definable using $\neg, \wedge, \vee$ as well as $\top, \bot$ and a new connective xx defined as follows:

$$
F_{\mathrm{xx}}(U, V) = \begin{cases} 1 & \text{if } U = V = 1 \\ 0 & \text{if } U = V = 0 \\ \ast & \text{otherwise.} \end{cases}
$$

But, there is a very natural combined notion too:

Proposition: The closed persistent truth functions are precisely those definable using $\{\neg, \wedge, \vee, \top, \bot\}$.

Proof: Let F be a closed persistent truth function. For each valuation V with $F(V) \neq 0$, define a formula β_V as follows:

Case 1: $F(V) = 1$. β_V is the conjunction of all p such that $V(p) = 1$ and all $\neg p$ such that $V(p) = 0$ (leaving out proposition letters receiving ∗).

Case 2: $F(V) = \ast$. β_V is the conjunction of Case 1 *plus* all conjunctions $q \wedge \neg q$ for proposition letters q receiving ∗. (As F is closed, at least one such q must occur.) There are some boundary cases: put $\top$ for an empty conjunction.

Now, the desired definition δ_F is the *disjunction* of all conjunctions β_V with $F(V) \neq 0$.

Claim: δ_F defines F.

- Let $F(V) = 1$. Since $V(\beta_V) = 1$, $V(\delta_F) = 1$, by the truth table for $\vee$.

- Let $V(\delta_F) = 1$. Then V gives value 1 to at least one disjunct $\beta_{V'}$. This cannot be a disjunct of the second kind (contradictions $q \wedge \neg q$ cannot receive value 1): and hence $F(V') = 1$, $V(\beta_{V'}) = 1$. It follows that $V' \sqsubseteq V$ (i.e., V is at most further defined than V', respecting its 0,1-decisions already taken), and hence by Persistence, $F(V) = 1$.

- Let $F(V) = 0$. Now, for no V' with $F(V') = 1$, V, V' can have a common $\sqsubseteq$-extension (using Persistence, and the incompatibility of values 0, 1). So, V, V' must disagree on at least one proposition letter where they are both non $*$. And therefore, $V(\beta_{V'}) = 0$. Thus, V refutes all disjuncts in δ_F which are of the first kind.

  Next, let $\beta_{V''}$ a disjunct in δ_F of the second kind. Again by Persistence, not $V \sqsubseteq V''$: and hence, either

  - V is defined on some proposition letter q where V'' is undefined—whence V falsifies the conjunct $q \wedge \neg q$, and consequently, $V(\beta_{V''}) = 0$, or

  - V, V'' disagree on some proposition letter where both are defined, and $V(\beta_{V''}) = 0$ for reasons similar to those above.

  So, in all, V refutes all disjuncts of δ_F: i.e., $V(\delta_F) = 0$.

- Finally, let $V(\delta_F) = 0$. That is, V conflicts with each valuation V' where $F(V') = 1$ on at least one proposition letter where both are defined (by the truth tables for $\wedge, \vee$). So, $F(V) \neq 1$. Moreover, for each V'' having $F(V'') = *$, $V''(\beta_{V''}) = *(!)$; whereas $V(\beta_{V''}) = 0$. So, V conflicts with every valuation where F is undefined too. It follows that $F(V)$ must be 0. ∎

There are many further questions which arise in this three-valued setting, generated by its potential for making new distinctions—such as the options for valid consequence mentioned before. A more extensive discussion of such issues may be found in J.-E. Fenstad, P.-K. Halvorsen, T. Langholm and J. van Benthem, 1987, *Situations, Language and Logic*, Reidel, Dordrecht and Boston. Many more technical results on persistence in partial predicate logic, and partial model theory generally, may be found in T. Langholm, 1987, *Partiality, Truth and Persistence*, dissertation, Department of Philosophy, Stanford University (to appear in CSLI Lecture Notes). In particular, the latter work is a mixture of new techniques, blended judiciously with sometimes surprising reductions to classical cases. (For instance, consequence in four-valued predicate logic reduces to ordinary consequence in predicate logic with only *positive* formulas.

One type of technical question concerning partial logic, however, deserves special attention, viz. its *complexity*. The move toward partiality is often motivated by an appeal to its greater attraction as a model of our actual cognitive activity. Implicit in this motivation is the idea that handling partial information is within human means, and can be done reasonably efficiently. But unfortunately, in the technical implementation of such ideas, a *paradox* threatens:

> more cognitively oriented logics often have a more complex notion of valid consequence than their classical counterparts!

An example is the case of intuitionistic versus classical inference (Chap. 7), or that of non-monotonic versus classical reasoning (Chap. 5). In this light, it is of interest to report a case of gain. At least, universal validity (though not valid consequence) in three-valued predicate logic is *decidable* (Langholm 1987), even when its classical counterpart is not. A general solution of the above paradox remains to be found, however, and may involve a much finer analysis of what kinds of statements (among all possible logical forms) are actually employed in our partial reasoning.

This concludes the present round of exercises in the logic of partiality. Many of these are still of a relatively traditional logical nature. More novel questions, generated by contemporary semantics of natural language, will be presented in the final Part of these notes.

IV   Recent Developments 3: Logical Semantics

This final part is devoted to some new research lines in Intensional Logic emanating from current work in the logical semantics of natural language. We shall be concerned mainly with so-called *generalized quantifiers*, and more generally, the *type theory* behind them.

A large class of natural language expressions may be viewed as denoting generalized quantifiers, and this notion proves a useful vehicle for further semantic theorizing. Two seminal papers in this development are J. Barwise and R. Cooper, Generalized Quantifiers and Natural Language, *Linguistics and Philosophy* 4, 1981, 159–219; and E. Keenan and Y. Stavi, A Semantic Characterization of Natural Language Determiners, *Linguistics and Philosophy* 9, 1986, 253–326 (but in circulation since 1981). A more extensive book on the resulting logical program is J. van Benthem, *Essays in Logical Semantics*, 1986, Reidel, Dordrecht and Boston.

A short survey of some highlights in this development will provide the necessary background for the next two chapters.

Determiner expressions in natural language combine with nouns to form noun phrases, which again combine with verb phrases to form verbs. Examples are *all, some, two, most* (simplex), or *not many, two or three, Cinderella's, no human, every beautiful farmer's daughter's* (complex). Extensionally, both common nouns and verb phrases denote sets of individuals (a wise initial idealization), and hence determiners may be regarded as denoting relations between sets of individuals. This may be pictured in the familiar Venn diagrams:

$D_E XY$ 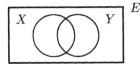 E E: universe of discourse
     X: CN-denotation
     Y: VP-denotation

For instance, *all* denotes inclusion, *some* overlap, etc.

Of the great multitude of possible denotations of this kind (with n elements in E, there are 2^{4^n} binary relations between its subsets), only a limited number seem to be realized by actual language expressions. To account for this, various *denotational* constraints have been formulated. Two major examples are the following.

Conservativity (CONS): The left-hand argument "sets the scene":

$$D_E XY \quad \text{iff} \quad D_E X(Y \cap X).$$

Example: All sinners burnt is equivalent to *All sinners were burning sinners:*

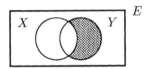

Topic-neutrality, or *Quantity* (QUANT): No individual entity plays a distinguished role:

$$D_E XY \text{ is determined completely by the numbers}$$
$$\#(X \cap Y), \ \#(X - Y), \ \#(Y - X), \ \#(E - (X \cup Y)).$$

Alternatively, for every bijection π defined on E,

$$D_E XY \quad \text{iff} \quad D_{\pi[E]}\pi[X]\pi[Y].$$

Example: All XY is equivalent to $\#(X - Y) = 0$, *most XY* to $\#(X \cap Y) > \#(X - Y)$.

Notice that CONS, QUANT together suppress the dependency upon $\#(Y - X)$. Actually, dependency upon the "contextual" $\#(E - (X \cup Y))$ is also rare, and we may choose to suppress it in a principle of

Context-Neutrality (EXT):

$$\text{if } X, Y \subseteq E \subseteq E', \text{ then } D_E XY \text{ iff } D_{E'} XY.$$

In addition to these general conditions, special purpose properties turn out to isolate important subclasses of determiners. Perhaps the best-known example is

Monotonicity: A determiner D is (upward) monotone if

$$D_E XY, \ Y \subseteq Y' \quad \text{only if} \quad D_E XY'.$$

Example: All birds fly, hence *All birds fly or walk*. Likewise for *some*, *most*. Downward monotonicity (in the obvious sense) occurs as well, e.g., with *no, few, at most three*.

When occurring in the left-hand argument, one often speaks of (upward) *persistence*:

$$D_E XY, \ X \subseteq X' \quad \text{only if} \quad D_E X'Y$$

(*some, at least two*), with a corresponding downward companion.

Monotonicity has been used extensively in linguistic and logical description. (Actually, it played already a crucial role in classical pre-Fregean logic.)

Another interesting classification of determiners is through well-known formal properties of binary relations. For instance, *all* (inclusion) is transitive, reflexive, anti-symmetric, *some* is (almost-) reflexive and symmetric, *not all* is irreflexive and strongly connected. These properties reflect central patterns of inference that can be supported by determiners.

Now, one interesting new phenomenon in this area has been the attempt to formulate "semantic universals," i.e., broad general laws concerning all human languages. Three examples are

- all determiners are conservative (Keenan and Stavi),

- all simplex determiners are upward or downward monotone, or a conjunction of the two kinds (Barwise and Cooper),

- no determiners are asymmetric.

The latter example is found in F. Zwarts, *Categorial Grammar and Algebraic Semantics*, dissertation, Nederlands Instituut, Rijksuniversiteit, Groningen, 1986. (To appear with Reidel, Dordrecht and Boston.)

Much subsequent logical research has been inspired by a desire to evaluate such universals. This is not so easy, especially since proposed theoretical conditions exert a definitional pressure on the earlier empirical classifications—a phenomenon not unknown from the history of "syntactic universals" in linguistics. For instance, it has been proposed to reject possessives and various other complexes to make all remaining "determiners" *logical* in the sense of QUANT.

In any case, logical results about these conditions are often *definability* theorems, matching an explicit set of linguistic descriptions with a certain set of denotations. This is still a form of "botany," of course, prior to a deeper understanding of semantic "laws." But, it is a first step towards semantic *theory*, beyond faithful transcription of natural language fragments into some semantic formalism.

One sign of the fruitfulness of the above theory is that it has turned out to be generalizable to many other types of expression than just determiners. A natural setting for describing such generalizations is a *type theory* in the style of Russell, Henkin, and Montague. We start with two basic types e ("entities" or individuals) and t ("truth values"), and then build up complex types by pairing. The pair (a, b) denotes all *functions* from a-type objects to b-type ones. Thus, one-place predicates of individuals have type (e, t), Noun Phrases type $((e, t), t)$, and determiners type $((e, t), ((e, t), t))$:

$$
\begin{array}{ccc}
\text{DET} & X & Y \\
((e,t),((e,t),t)) & (e,t) & \\
\hline
((e,t),t) & & (e,t) \\
\hline
t &
\end{array}
$$

(Of course, in our relational analysis, we have "flattened" the determiner type somewhat, to bring both arguments on a par.) Other important types are, e.g., (t, t) (operators on truth value expressions, i.e., unary sentence operators), $((e, t), (e, t))$ (operators on unary predicates, such as adverbs or adjectives). A systematic account of this type-theoretic background for generalized quantifiers may be found in the earlier-mentioned book *Essays in Logical Semantics*.

Interestingly, the above advances have been made largely by *disregarding* intensional phenomena—obsession with which has tended to obscure the many insights still to be discovered about "elementary English." Nevertheless, there is no obstacle in principle towards asking the same questions about intensional notions, trying to understand the basic linguistic modes of expression in this realm through a motivated set of semantic constraints on all a priori denotations. In the following chapters, we shall return to earlier topics in this light.

The proper type-theoretic setting for this investigation employs not just basic types e, t, but also s (possible worlds, or more generally "indices of evaluation"). (See R. Montague, *Formal Philosophy*, 1974, Yale University Press, New Haven (R. Thomason, ed.), and also D. Gallin, *Intensional and Higher-Order Modal Logic*, 1975, North-Holland, Amsterdam.) A proposition will then be an object of type (s, t), assigning a truth value to each world (point in time, possibility, ...). Thus, the modal and tense operators of Chapters 1 and 2 come out in type $((s, t), (s, t))$, to be studied in Chapter 9. On the other hand, conditionals *if* come out more like the above determiners, viz. in type $((s, t), ((s, t), (s, t)))$, which may be rearranged to $(s, ((s, t), ((s, t), t)))$ and then, suppressing one index dependence, to $((s, t), ((s, t), t))$. The latter will be investigated in Chapter 10. The final chapter then raises some general issues of Intensional Type Theory.

9   Denotational Constraints on Tense and Modality

Literature

Van Benthem, J. 1986. Tenses in Real Time. *Zeitschrift für mathematische Logik und Grundlagen der Mathematik* 32, 61–72.

Van Benthem, J. 1986. A Linguistic Turn: New Directions in Logic. In R. Marcus et al., eds., *Proceedings 7ʰ International Congress of Logic, Methodology and Philosophy of Science. Salzburg 1983*, Amsterdam: North-Holland, 205–240.

There is only a limited number of tenses in natural language, and we want to see how this phenomenon can be understood in terms of denotational constraints.

Tenses in Real Time

Let us fix one temporal structure, the real number line $\mathbf{R}$. Tenses may be regarded as operations on propositions, that is, in the model, as functions taking sets of real numbers to other such sets. This time, we are interested in a structural analysis of the latter perspective, rather than the earlier linguistic ones (such as Kamp's equation of "tense" with "first-order definable schema").

A first, obvious condition is that tenses respect the temporal order. Technically, this amounts to invariance for order-automorphisms (for a tense f): (AUT)

> For all order-preserving permutations π of $\mathbf{R}$,
> $\pi[f(X)] = f(\pi[X])$, for all $X \subseteq \mathbf{R}$;

or, equivalently,

> $y \in f(X)$ iff $\pi(y) \in f(\pi[X])$, for all $y \in \mathbf{R},\ X \subseteq \mathbf{R}$.

93

(As to the importance of order automorphisms in our thinking about time, compare the earlier discussions of *Homogeneity* in Chaps. 1, 6.)

All first-order definable tenses obey AUT; but the converse is not true.

Against this background, an attractive further condition is suggested by the behavior of the Priorean tenses, which are all *continuous* in the following sense: (CONT)

$$F\left(\bigcup_i X_i\right) = \bigcup_i f(X_i), \text{ for all families } \{X_i \mid i \in I\}, X_i \subseteq \mathbf{R}.$$

These two conditions contain the essence of the Priorean tenses in Chapter 1:

Theorem: The only **R**-tenses satisfying AUT and CONT are those defined by the schema "f is some union of *pa*, *pr*, *fu*," where

$$pa(X) = \{y \in \mathbf{R} \mid \exists x \in X \ x < y\},$$
$$pr(X) = X,$$
$$fu(X) = \{y \in \mathbf{R} \mid \exists x \in X \ y < x\}.$$

Proof: First show, using AUT, that f has a limited range of choices for each singleton set $\{x\}$. Its image either contains or is disjoint from each of the three regions $\{y \in \mathbf{R} \mid y < x\}$, $\{x\}$, and $\{y \in \mathbf{R} \mid x < y\}$. Moreover, f makes the same choice for all $x \in \mathbf{R}$, again by AUT. Thus AUT enforces a strong uniformity of behavior.

Next, CONT says that these pointwise choices determine the behavior of f completely, as $f(X) = \bigcup_{x \in X} f(\{x\})$. ∎

On the other hand, the *progressive* tense fails the CONT test: it gives the *empty* set for each singleton, and yet it assigns non-empty sets to, say, open intervals. (In our structural perspective, the progressive just denotes the topological *interior operation*.) Still, a relaxed version of the idea behind CONT ("local choices") remains valid here: (BC)

$$y \in f(X) \text{ iff } y \in f(X_0), \text{ for some *bounded convex* episode } X_0 \text{ in } X.$$

A more complicated combinatorial version of the above argument will classify all AUT, BC tenses; yielding, in addition to the Priorean tenses and the progressive, some "tenses" true only in certain *boundary points*. Thus, a hierarchy of tenses (or more generally, operators of *temporal perspective*) arises, with ever weaker constraints added to AUT; for which a natural end point seems to be the earlier *monotonicity*:

$$\text{if } X \subseteq Y, \text{ then } f(X) \subseteq f(Y).$$

By this time, infinitely many tenses will qualify.

There are several directions to go from here. For instance, one can do the same type of analysis on other temporal structures, such as the integers **Z**, which have fewer automorphisms than **R**, and hence tolerate more AUT, CONT tenses. But perhaps the most significant question is the following. Once we have this restricted perspective upon tenses, as special classes of operations from propositions to propositions, it seems reasonable to extend it to the propositions themselves. (Compare a similar observation made in Chap. 2.) After all, the more esoteric mathematically possible subsets of **R** do not seem to qualify as the "lifetimes" of natural language propositions. (Compare also the Prior-Hamblin ban on "indefinitely finely intermingled" truth values 0,1, mentioned in Chap. 6.) One reasonable limited range would be that of *convex* (uninterrupted) sets of reals, these being the lifetimes of ordinary events—or, at most, finite unions of these (allowing "repetitive events"). In such a setting, much of the correspondence and axiomatizability theory in Chapter 1 will have to be rethought.

A Structural View of Modality

Somewhat more abstract, but equally feasible is the view of modalities as structural operations on propositions. In a very simple setting, one can think of mere sets of possible worlds, as yet without a relational pattern. In fact, one would like to investigate just when one is forced to introduce the latter.

A general constraint (as above) is "world-neutrality" (QUANT):

$$\pi[f(X)] = f(\pi[X]), \quad \text{for all permutations } \pi \text{ of the universe } W$$
$$\text{of possible worlds.}$$

As a consequence, for any set $X \subseteq W$, $f(X)$ will either contain or avoid X (as well as $W - X$) in its entirety. (The reason is this. If $f(X)$ contains $x \in X$, and y is another object in X, then the interchange of x, y (with all other worlds undisturbed) is a permutation π of W leaving X in its place. But, then, $x \in f(X)$ implies $\pi(x) \in \pi[f(X)] = f(\pi[X]) = f(X)$: and hence $y \in f(X)$ too.)

On this basis, modal logics may be analyzed; for instance, the logic $T = K + \Box\varphi \to \varphi$. We restrict attention to *finite* models.

Theorem: The only QUANT T-modalities are *identity* ($f(X) = X$: i.e., $\Box$ collapses into truth) and *S5-necessity* (i.e., $f(X) = W$ if $X = W$, $f(X) = \emptyset$, otherwise).

Proof: First, by the above observation on the effect of QUANT, for any set X, $f(X)$ must equal $\emptyset$ or X or $W - X$ or W. Moreover, by the special T-axiom, $f(X) \subseteq X$—and so only the first two possibilities remain. Now, since K has the principle $\Box$-*true*, it follows that $f(W) = W$. Then if, in addition, $f(X) = X$ for all X, we are in the above first case. If, on the other hand, $f(X) \neq X$ for some X (i.e., $f(X) = \emptyset$, while $X \neq \emptyset$), then the following must be true. Let

X be of *maximal* size such that this phenomenon occurs. [Notice that $\#(X) < \#(W)$). In fact, $\#(X)$ must be $\#(W) - 1$. (For, otherwise, two worlds w_1, w_2 could be picked in $W - X$, leading to the following contradiction : $f(X \cup \{w_1\}) = X \cup \{w_1\}$, $f(X \cup \{w_2\}) = X \cup \{w_2\}$, $X = (X \cup \{w_1\}) \cap (X \cup \{w_2\})$, and yet $f(X)\,(= \emptyset) \neq f(X \cup \{w_1\}) \cap f(X \cup \{w_2\}) = X\,(\neq \emptyset)$. The latter conclusion contradicts the basic K-axiom $\Box\varphi \wedge \Box\psi \to \Box(\varphi \wedge \psi)$.] By QUANT then, $f(X) = \emptyset$ for *all* X of size $\#(W) - 1$. Then finally, by the monotonicity principle $\Box(\varphi \wedge \psi) \to \Box\varphi$ of K, $f(X)$ will be empty for all proper subsets X of W—and $S5$-necessity results. ∎

With the minimal modal logic K only, a few more possibilities remain. Even so, QUANT leaves too few modal operations for comfort, and we shall drop it. This means allowing, in general, that $f(X)$ can indeed be sensitive to differences between individual worlds: some are more "prominent" than others. One appropriate measure of this is by a relational hierarchy, and we are witnessing the birth of the earlier alternative relation R.

One way of introducing R occurs in the modal folklore.

Theorem: With a strong reading of the above K-principles as

$$f\left(\bigcap_i X_i\right) = \bigcap_i f(X_i), \text{ for all families } \{\, X_i \mid i \in I \,\}, \; X_i \subseteq W,$$

a function f from $P(W)$ to $P(W)$ satisfies the minimal modal logic if and only if there exists some binary relation R on W such that

$$f(X) = \{\, w \in W \mid \forall x(Rwx \to x \in X)\,\}.$$

Thus, semantic models $\langle W, R \rangle$ will come to the fore; and one is left only with a requirement of *Quality* (compare the earlier AUT):

$$\pi[f(X)] = f(\pi[X]), \text{ for all } R\text{-}automorphisms \; \pi \text{ of } W.$$

Again, definability issues like those for tenses may be raised here.

Another angle is to return to the linguistic perspective on modalities, viz. all possible first-order definitions for $\Box$ in terms of R. Which proposals in this spectrum of "truth definitions" satisfy the conditions of the minimal modal logic? (The idea of a diversity of truth definitions cannot be too great a surprise, after the variety mentioned in Chap. 3.) It may be shown that only one kind of candidate qualifies, viz. the "Kripke family"

$$f(X) = \{\, w \in W \mid \forall x \in W(\rho(w, x) \to x \in X)\,\}$$

where ρ is some formula in $R, =$ only. For a proof, see J. van Benthem, *Possible Worlds Semantics: a research program that cannot fail?*, *Studia Logica* 43:4, 1984, 379–393.

Finally, it would be of interest to carry out a similar analysis of denotational constraints on propositional operators in structures of *partial situations*. As these come in a natural pattern of *inclusion*, such operators will have to respect inclusion automorphisms on situations. For a first discussion, see J. van Benthem, 1987, *Categorial Grammar and Type Theory*, report 87–07, Institute for Language, Logic and Information, University of Amsterdam (to appear in *Linguistics and Philosophy*).

Temporal Conditionals

The preceding modal intermezzo raises the question whether the emergence of a temporal precedence order can also be explained, rather than taken for granted. One possible road here is the study of the temporal conditional *if* φ, *then* (subsequently) ψ. With suitable constraints imposed on its denotation, as in the earlier-mentioned case of determiners, this conditional induces a precedence order $<$ as follows:

$$x < y \quad \text{iff} \quad [\![if]\!]\{x\}\{y\}.$$

Further details are provided in the above-mentioned Salzburg paper. Actually, this case is treated there as a corollary to a method for extracting similarity hierarchies from ordinary conditionals (compare Chap. 3):

$$C_x yz \quad \text{iff} \quad [\![if_x]\!]\{y, z\}\{y\}.$$

In the following chapter, however, a more thorough study will be made of the latter area.

For the moment, it has been illustrated how the present perspective allows us to start thinking systematically about the genesis of our intensional model-theoretic apparatus, and the available options. This theme will be elaborated in what follows.

10 Conditionals as Generalized Quantifiers

Literature

Van Benthem, J. 1984. Foundations of Conditional Logic. *Journal of Philosophical Logic*, 13:3, 303–349.

Veltman, F. 1985. *Logics for Conditionals*. Ph.D. dissertation, Filosofisch Instituut, University of Amsterdam. Forthcoming Cambridge University Press.

Conditionals may be viewed as denoting generalized quantifier relations between sets of possible worlds, viz. the denotations of their antecedent and consequent propositions. Intuitively, "enough" of the former should be included among the latter—and different conditional relations may be more or less tolerant of exceptions. Thus, conditionals are treated here as (meta-) relations among propositions, rather than (object) operations upon these—although we shall try to have our cake and eat it. This rich and useful confusion has a long-standing logical tradition.

The Basic Framework

As with determiners, the basic picture for conditionals is given in a Venn Diagram:

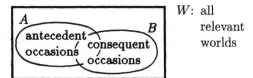

W: all relevant worlds

Examples: *all A* are *B, all but three, almost all, most, . . .*

99

Again, we search for general constraints. Rather than pointing at specific "valid" and "invalid" argument schemata for *if* (a contested area in any case, witness Chaps. 3, 7), we want *general intuitions* on what makes a binary set relation "conditional." This preference is just one instance of a general tendency advocated in these notes, beginning with the earlier interest in "global intuitions" on temporal models (Chap. 1). Indeed, our faculty of "intuition" (if it exists) would seem to apply more properly in the latter area than in very concrete endorsements of specific inferences.

Very typical for conditional relations is the idea of CONFIRMATION. Shifting the balance between "confirming instances" $(A \cap B)$ and "exceptions" $(A - B)$ in favor of the former, leaves a conditional true. This statement translates into various clauses:

- $if_W AB$ implies $if_W A(B \cup C)$ (right monotonicity),

- $if_W AB$ implies $if_W (A \cup C)(B \cup C)$ (addition of new confirming instances),

- $if_W A(B \cap C)$ implies $if_W (A \cap B)C$ (removal of exceptions).

Notice that the latter clause is that axiom in the minimal conditional logic of Chapter 3 which replaces classical left-monotonicity:

$$if_W AC \quad \text{implies} \quad if_W (A \cap B)C.$$

The latter principle seems unappealing in general: in decreasing A, we might be removing confirming instances.

Then, as with determiners in general, the antecedent argument sets the scene for the conditional:

ANTECEDENCE: $If_W AB$ iff $if_W A(B \cap A)$.

The next constraints are of an even more general logical nature. (This is appropriate, since we want to think of conditionals as "logical" relations in some sense.) Again, one of these has been mentioned before, viz. *context-neutrality*:

EXTENSION: When $A, B \subseteq W \subseteq W'$, $if_W AB$ iff $if_{W'} AB$.

The following two ideas have not yet been encountered.

ACTIVITY: A logical constant should do some work; in particular, *if* should depend on both arguments. We stipulate:

For each non-empty set A, there exist $B, B' \subseteq A$ such that
$if_W AB$, not $if_W AB'$.

This condition rules out quite a few cases—but it is very convenient for getting at essentials.

More difficult to formulate, but perhaps most intriguing of all is our feeling of UNIFORMITY. The (truth value) behavior of a conditional should be the same everywhere. One way of making this precise is by considering the following "thought-experiment":

old situation: A, B

add exception: A_+, B add confirming instance: A^+, B^+

add both: A_+^+, B^+.

Of the sixteen possible patterns here, Confirmation and Activity rule out ten, and we are left with

$$
\begin{array}{ccccccc}
 & T & & T & & T & & F & & F & & F \\
T & T & F & T & F & T & F & F & F & T & F & T \\
 & T & & T & & F & & F & & F & & T
\end{array} .
$$

Now Uniformity says that, at least once the two separate experiments have been performed with a value for their combination, this outcome will be stable. Thus, for example, the second and third patterns could not both occur for one and the same conditional relation.

Actually, there is a whole *hierarchy* of Uniformity constraints, a topic not to be pursued here. (See *Essays in Logical Semantics* for a more general treatment, involving various types of *automata* representing uniform procedures for computing truth values.)

Limitations of the Barest Setting

In the most austere case, conditional relations can only involve what is *explicitly* given by their linguistic context: that is, the two bare sets of antecedent and consequent worlds, without any "hidden variables." What this amounts to is the earlier condition of *Quantity*. Here, we shall not regard this as the hallmark of "logicality" (as in the case of determiners), but rather as the fundamental stage of least ontological commitment. If one has to part company with Occam, it had better be for good reasons.

In addition, attention will be restricted to *Finite Universes W*. Recall that this was the area where traditional conditional semantics ran most smoothly (Chap. 3). Again, the matter may also be viewed as one of commitment: in semantics, we only want to consider matters of infinity if there are clear reasons for doing so.

In this austere realm, we run up against an instructive trilemma.

Theorem: The only conditionals satisfying all conditions specified above are *all, half or more*, and *some*.

Proof: First, conditional relations satisfying QUANT, CONS, EXT may be represented on the finite models as subsets of the following 'Tree of Numbers':

$$
\begin{array}{ccccc}
|A| = 0: & & & 0,0 & \\
1: & & 1,0 & & 0,1 \\
2: & & 2,0 & 1,1 & & 0,2 \\
3: \; 3,0 & & 2,1 & & 1,2 & & 0,3 \\
& & & etc. & \\
\end{array}
$$

Here, pairs i, j indicate numerical contents for the sets of exceptions $(A - B)$ and confirming instances $(A \cap B)$, respectively. The conditional is specified completely by marking which distributions it accepts (or rejects).

The use of this representation (here, and elsewhere) is that further constraints now translate into simple *geometrical conditions* on the above "acceptance sets." For instance, Activity demands that every horizontal row, except the first, have both accepted and rejected positions. Confirmation says that, if a position is accepted, then so is every position to its right, its southeast and its northeast (following the three clauses in the order given above). With these insights, a simple argument about the possibilities left open by Uniformity will yield the conclusion. ∎

A Map of Ways-Out

Evidently, the above three conditionals are too small a set for modelling our various notions of conditionality. For instance, none of them validates precisely the minimal conditional logic of Chap. 3. Therefore, the escape routes from the trilemma become of interest. These may be charted systematically, by dropping one or the other of the intuitions and assumptions going into the above proof. One obvious possibility here is to give up the assumption made last, that of finiteness, and to investigate additional possibilities on *infinite universes*. As it turns out, few intuitively attractive candidates arise that way. The next most obvious move is to give up Quantity, in favor of some "hidden variable hypothesis." The two major directions here are either *inductive*, assigning different "weights" to the various worlds by means of some probability measure, or *intensional*, introducing some hierarchy of relative prominence among worlds (as in Chap. 8). Along the latter line, of course, we are witnessing the genesis of the traditional approach of Chapter 3—about which one may again ask various questions of definability, since most of the above constraints are still in force.

Inverse Logic

The emphasis on broad intuitions, rather than specific inferences, also leads to a more detached "reverse" perspective upon logical description of the latter. Against the background of our general constraints on conditional relations, one may study "realizable" logics. For instance, a result in J. van Benthem, 1983, Determiners and Logic, *Linguistics and*

Philosophy 6, 447–478, implies that no conditional validates logics with the pattern

$$\text{if } AB, \text{ if } BC \quad \text{imply} \quad \text{if } CA.$$

Thus, the popular practice of circular reasoning has no hope for a foundation in logic.

An example of a result from the above-mentioned conditionals paper will illustrate the above general trend. It concerns the well-known *Scott Conditions* on logical inference: reflexivity, transitivity and (left-and-right) monotonicity.

Observation: The first two Scott conditions imply the third.

Theorem (Antecedence, Activity): The only conditional satisfying the Scott conditions is classical entailment, that is, *all A are B*.

Actually, the proof shows that *left-monotonicity* is already virtually equivalent to demanding this one classical conditional; which explains the "non-monotonicity" found in so many newer theories of conditionality. (Compare Chap. 5.)

Many other examples of this kind of analysis of proposed inferential patterns could be added.

This concludes Chapter 10. Much attention has been devoted to conditionals in this fourth part of our notes: not only for their intrinsic interest, but also because this simple case study exhibits many notions and ways of thinking of a much wider semantic applicability. For instance, a similar study of proposed notions of *Verisimilitude* in the philosophy of science may be found in J. van Benthem, 1987, Verisimilitude and Conditionals, in T. Kuipers, ed., *What is Closer-to-the-Truth?*, Rodopi, Amsterdam, 103–128. But also, one could give an analysis of *conditional instructions* in programming languages, using our type of denotational analysis (cf. the earlier-mentioned paper on *Categorial Grammar and Type Theory*). Thus, we hope to have imparted a useful *perspective*, rather than a mere set of specific results.

11 Intensions and Types

At various places in the preceding chapters, types were attached to expressions, thus assigning the latter a locus in a more general hierarchy of abstract semantic objects. But, is there any intrinsic merit to this conjunction of Intensional Logic and Type Theory? For a start, there is a *practical* need. If we are to have a logic powerful enough to describe significant parts of natural language, we will have to incorporate both intensional and type structure. And in fact, this is just what Richard Montague did in his pioneering work around 1970. Still, the resulting *theory* is somewhat disappointing from a logical point of view; witness Gallin's book *Intensional and Higher-Order Modal Logic*. The logical results obtained there are by and large the expected generalizations from the separate components, wrapped in layers of "cups" and "caps." And even the latter vanish, once one takes a full type theory with primitives e, t, s (Gallin's "Ty 2"), which obviates the need for special intensional provisos on axioms and rules.

Type Change

Nevertheless, more interesting questions do arise, precisely because logical semantics has brought to light such interesting structures in the *extensional* realm as generalized quantifiers and their ilk. For, at least, this raises the question how insights obtained in this way may be preserved in the transition to a full intensional setting. (See E. Keenan and L. Faltz, 1985, *Boolean Semantics for Natural Language*, Reidel, Dordrecht and Boston, on this transition for the case of Boolean structure in natural language.) On the other hand, there has also been a "downward" tendency among Montagovians these days. Faced with the formidable complexity of the full intensional type theory as it applies to natural language expressions, they have tried to find a systematic "deconstruction" of Montague's intensional types, as arising from more obvious simple cases by certain systematic rules of *type change*. (See M. Rooth and B. Partee, 1983, Generalized Conjunction and Type Ambiguity, in R. Bäuerle et al., eds., *Meaning, Use and Interpretation of Language*, De Gruyter, Berlin, 361–383.)

Actually, one should distinguish at least two forms of type change: both of which play a role in Montague's system. One has to do with the general phenomenon of *polymorphism* in natural language, which already occurs with purely extensional types constructed from e and t. For instance, negation can occur not only on full sentences, but also on verbs ("not love"), noun phrases ("not every minute"), adverbs ("not voluntarily"), etc. Thus, in addition to the basic (t, t), one would also need $((e, t), (e, t))$, $(((e, t), t), ((e, t), t))$, $(((e, t), (e, t)), ((e, t), (e, t)))$, etc. Again, an unmanageable complexity threatens. But, the following simple rule of type change (proposed by Peter Geach) proves all that is needed:

> If an expression can occur in type (a, b), then it can also occur in type $(((c, a), (c, b))$, for arbitrary c.

Moreover, this rule already removes one of the complexities in Montague's system, viz. his treatment of transitive verbs. The natural type of the latter is just $(e, (e, t))$. But, he felt compelled to reject this, because of the difficulty in parsing a sequence like "love every minute":

$$(e, (e, t)) \quad ((e, t), t)$$

does not combine by function application, either way. Therefore, Montague had to treat transitive verbs as being of type $(((e, t), t), (e, t))$: a decision with several combinatorial repercussions elsewhere. But, by the Geach Rule, one simple step suffices to make the preceding example fit without further ado:

$$((e, t), t) \text{ changes to } ((e, (e, t)), (e, t)).$$

By now, there is an extensive theory of this kind of type change, and its logical properties, for which we must refer to the literature. (See, e.g., *Essays in Logical Semantics*, Chap. 7.)

Strategies of Intensionalization

Of more interest to us here is truly intensional type change. We have a reasonable picture of the behavior of extensional types of expression. Are there *systematic* ways in which intensionality can be introduced here? For instance, a look at any table of intensional types for propositions, verbs, etc., suggests strong similarities, which may be summed up as follows ("Montague's Strategy"):

a. replace former t by (s,t),

b. replace former e by (s,e).

Here, we shall only consider a possible systematic reason for the first of these. (The second is more controversial in any case, as we shall see.)

The preceding discussion may have given the wrong impression that introducing a possible worlds perspective (with type s) is a momentous and controversial move, forced upon us only by the consideration of exotic constructions in natural language. But actually, some of our previous points of view at least suggest making this move for entirely natural reasons.

Consider the extensional e, t-grammar as before. Objects of type e were to be *entities*. What about those of type t? It would seem reasonable not to prejudge issues here by unduly restricting these, say to just $\{0, 1\}$. Instead, one can think of type t objects as *propositions* in some intuitive sense. Moreover, the latter come as an ordered structure, with a relation of implication, or even a full Boolean Algebra. In view of the actual structure of sentences, the latter seems a realistic assumption. Now, the usual ultrafilter technique for Boolean Algebras will represent the class of propositions as an algebra of subsets of some carrier set S of ultrafilters (or "maximal state descriptions"). (The procedure is analogous to the Henkin model construction, found in Chap. 8.) Letting these states be the objects of a new type s, one arrives at an e, t, s-grammar, standing for a "Boolean-valued" version of the original one.

This introduction of s gives us "possible worlds." In addition, Chapter 9 provides an equally harmless emergence for the "alternative relation." For, once it is admitted that there exists a propositional operator $\Box$ satisfying the postulates of the minimal modal logic (and this much seems quite uncontroversial), a suitable relation R may be introduced on S as in the proof of the last theorem in Chapter 9. Thus, the $\langle W, R \rangle$-perspective of the first part of these notes is ubiquitous, and hence not overly restrictive.

Still, this reconstruction produces *total* worlds in the traditional sense. Why has no partiality arisen? The latter phenomenon would occur if one assumes less structure for the propositions, say as a mere *partial order* ("implication") with greatest lower bounds ("conjunctions"), again admitting a necessity operator as above. This time, a *filter* representation will work (rather than an ultrafilter one), producing a structure $\langle W, \sqsubseteq, R \rangle$ much like Humberstone's possibility models (see Chap. 7).

With the reinterpretation of the propositional type t to (s,t), where the new t now really stands for truth values, the original *predicates* (e, t) become of type $(e, (s,t))$. (In other words, a predicate assigns a proposition to each entity.) Technically, this is equivalent to $(s, (e, t))$, with the argument order reversed—that is, the usual multiple reference way of thinking about predicate denotations: extensions along all possible worlds. With more complex types, we get ever more intensionality. For instance, the noun phrase type $((e,t),t)$ goes, essentially, to $(s,((s,(e,t)),t))$. This would be our reason for leaving the type e alone, as this observation by itself already solves the "temperature puzzle" which made Montague introduce his *intensional concepts* of type (s,e).

Intensional versus Extensional

The move towards an intensional type theory also raises various more theoretical questions. For instance, one conspicuous phenomenon is that certain expressions remain "extensional," even in their new intensional guise. One example is *negation*, whose behavior on propositions is as follows:

$$\lambda P . \lambda s . \neg P(s).$$

That is, the old truth-functional negation is used ("pointwise") at each state, and these results are then collected. Thus, there arises a general question, pictured in the following diagram:

$$
\begin{array}{ccccccc}
& s & & s & & s & s \\
P & \downarrow \rightsquigarrow & \downarrow & A(P) & P \downarrow & & \downarrow \; A^* \circ P \\
& (e,t) & & (e,t) & & (e,t) \rightsquigarrow (e,t) & \\
& & & & & A^* &
\end{array}
$$

When can a function A from $(s,(e,t))$ to $(s,(e,t))$ be represented by a function A^* from (e,t) to (e,t) in the sense that

$$A(P)(s) = A^* \circ P(s)?$$

There is an answer here in terms of suitable conditions of Quantity and Extension. But, the straightforward set-theoretic criterion is this:

$$
\text{for all } P, Q \text{ and all } x, y \text{ in } S: \\
\text{if } P(x) = Q(y), \text{ then } A(P)(x) = A(Q)(y).
$$

Interestingly, essentially the same question arises from a suggestion of David Kaplan's at the 1984 conference at Palo Alto in his honor. He proposed admitting a "relational" adverb *necessarily*, connecting an individual and a property, with the task of showing subsequently how such a usage could be "reduced" to the original sentential one. In our types, adverbial $\square$ has $(e,((e,t),t))$, or equivalently $((e,t),(e,t))$—which becomes $(e,(s,t)),(e,(s,t))$ after the intensionalisation. But then, Kaplan's task is to suppress the dependence upon e, just as the above suppressed the dependence upon s.

Actually, the "minimal" meaning shift involved in the above examples is very reminiscent of those which may be associated with the earlier more combinatorial type changes. For instance, the intensional type

$((s,t),(s,t))$ for negation would also arise from the original (t,t) via the Geach Rule. We will not pursue this connection here.

One general issue raised by the preceding discussion is a rather surprising question. What do we *mean* by *Extensionality* in the general setting of an intensional type theory? Of course, there are some standard examples in simple cases, whose common feature is a certain "locality" when evaluating in a certain world. But curiously, no general satisfactory definition seems to exist.

General Questions

Then, there is the question of what becomes of the earlier logical themes in a general type-theoretic setting. We have seen how certain *denotational constraints* made sense, both in extensional and intensional settings. Can these be generalized to arbitrary intensional types? As it happens, the answer is positive for several notions, such as *permutation invariance* or *monotonicity*. (For details, see *Essays in Logical Semantics*.) One interesting feature of such generalizations is that we see new similarities across diverse domains of applied logic. Here is one example.

The *Dynamic Logic* of Chapter 5 views deterministic programs as functions from states to states of some computer: type (s,s). Then, basic program constructions act as higher operators. E.g., Composition (;) has type

$$((s,s),((s,s),(s,s))),$$

and Iteration (WHILE ... DO ...) has type

$$((s,t),((s,s),(s,s)).$$

Interestingly, these operations are both permutation-invariant (in the appropriate generalized sense) in their respective domains. Thus, one can analyze logical notions of *control* across different types in terms of this basic feature. And this remains possible, even if we change to a view of programs as denoting transition relations on states: type $(s,(s,t))$. Likewise, it makes sense to talk about general *monotone* operators here (cf. the earlier-mentioned paper *Categorial Grammar and Type Theory*).

To be more concrete, here is one illustration showing how the earlier analysis of denotational constraints in Chapter 9 makes sense now as well. Some basic operations on program transition relations found in the literature are *conversion*, *diagonalization* ($\triangle(R) = \{(x,x)|(x,x) \in R\}$), *composition*, *union* and *intersection*. These operations all have the following properties, formulated earlier. They are *permutation-invariant*, in that they commute with arbitrary permutations of individuals, and they are *continuous*, in that they commute with arbitrary unions of argument sets. By extending the analysis of Chapter 9, all such operations are completely determined by their behavior on singleton relations consisting of one ordered pair. It is not hard to enumerate all possibilities here: there are only finitely many of them. (See *Categorial Grammar and Type Theory*, Appendix 7, for the special case of unary operations.) A normal form of definition will be [for non-empty relational arguments]

$$\lambda x_s y_s \; . \; \lambda P_{s,(s,t)} Q_{s,(s,t)} \; . \; \exists z_s u_s \in P \exists v_s w_s \in Q \; . \; followed\ by$$

"Boolean combination of identities involving x, y, z, u, v, w."

E.g.,

$$\text{conv}(P) = \lambda xy \; . \; \lambda P \; . \; \exists zu \in P \; . \; x = u \wedge y = z,$$
$$\text{comp}(P, Q) = \lambda xy \; . \; \lambda PQ \; . \; \exists zu \in P \exists vw \in Q \; . \; u = v \wedge z = x \wedge w = y,$$
$$\text{un}(P, Q) = \lambda xy \; . \; \lambda PQ \; . \; \exists zu \in P \exists vw \in Q \; . $$
$$(x = z \wedge y = u) \vee (x = v \wedge y = w).$$

In fact, the above examples are distinguished among these forms by some further special conditions, making them "positive" and restricted to individuals occurring in the arguments P, Q. These conditions will not be spelled out formally here.

The point of this excursion has been to show that the earlier analysis of denotational constraints for intensional operators, and the resulting hierarchy of mathematical behavior, carries over to this new domain of application. Thus, the type-theoretic perspective draws together various strands from this Manual.

Nevertheless, there is no claim here that Montague's type theory is the only possible way of embedding Intensional Logic into a more general logical and linguistic background theory. Already, the development in Part III showed that, at least, suitable *partialized* versions of the enterprise will have to be considered. (For an example, compare R. Muskens, 1987, *Going Partial* and *Going Relational in Type Theory*, Filosofisch Instituut, University of Amsterdam.) And in the long run, other general semantic frameworks may turn out to be preferable.

Even the present tentative exploration may have shown, however, that it makes sense to embed Intensional Logic in a larger logical enterprise.

Appendix: Partial Type Theory

If Intensional Logic needs an infusion of partiality (as was suggested in Part III) as well as an embedding in the above environment of higher types, various earlier questions will have to be rethought in such a setting. For instance, we need suitably generalized formulations of the central phenomena in Chapter 8. Notably, what is *persistence* in higher types? Just to illustrate a possible road, here is a definition for a relation $\leq$ of "being more specific" on arbitrary type domains:

- $\leq_t$ is given by the familiar relation $*\ \begin{smallmatrix} 0 \\ 1 \end{smallmatrix}$,

- $\leq_e$ is ordinary identity of individuals,

- $\leq_{(a,b)}$ compares functions as follows:

$$f \leq_{a,b} g \text{ if, for all } x \in D_a, f(x) \leq_b g(x).$$

This will lead to intuitively correct outcomes on, e.g., partial predicates of individuals. Now, any expression τ containing some parameter u may be *persistent* with regard to u, in the sense that replacing the denotation of u by some more specific item will only make the denotation of τ more specific. In Chapter 8, we saw the special case of this where τ was a sentence (of type t) and u some subsentence. In partial predicate logic, the same would happen for sentences with respect to predicate parameters (of types (e,t), $(e,(e,t,))$, etc.).

Question: To prove a general syntactic characterization of persistence in a type-theoretical language set up using both negations $\neg$ and $\sim$ as well as the usual notions of application and lambda abstraction.

To be more specific, we consider what the simplest type-theoretic level beyond first-order predicates is, being unary *generalized quantifiers*

$$Qx \,.\, \varphi(x)$$

of type $((e,t),t)$. Our task is now to provide a partialized version of the semantic considerations in Chapter 9 and 10. Here are some observations, showing that this is quite feasible.

1. Partial generalized quantifiers Q assign values $1,0$ or $*$ to situations of the following kind:

$$A^+ \quad A^* \quad A^-$$

2. Earlier *general constraints* carry over, such as *permutation invariance*, which will now make Q sensitive only to the cardinalities of the three zones. Again, a useful numerical representation results in a 'Tree of Numbers'. For a universe of individuals of size n, the numerical space is as follows:

		$|A^+|, |A^-|$		
$|A^*| = n$		$0,0$		
$n-1$		$1,0$	$0,1$	
.				
.				
.				
0	$n,0$	$n-1,1$	$1,n-1$	$0,n$

Examples of truth value patterns for quantifiers may now be depicted geometrically (e.g., with $n = 4$):

all	some	at least two
*	*	*
*−	+*	**
* − −	+ + *	+ * *
* − −−	+ + +*	+ + *−
+ − − − −	+ + + + −	+ + + − −

3. These patterns actually demonstrate some further reasonable conditions, similar to those encountered for three-valued logic in Chapter 7.

> *Closedness.* The final row of total distributions for A^+, A^- over the universe should contain no values $*$.

> *Prediction.* If both extensions $x + 1, y$ and $x, y + 1$ of some position x, y have the same value, then x, y itself should already display that value.

> *Persistence.* Truth values 1 or 0 should propagate downward in the Tree.

It is easily shown that, given any classical "total" quantifier Q, these conditions leave just one unique extension Q_p to all partial situations. (Form the "minimal triangles" on the top of the base line.)

But of course, one might experiment with weaker conditions (allowing certain *variants* of *all* or *some*), or *different* conditions, inspired by further themes in Chapter 9 and 10.

4. There is a special place here for a first-order logic with the earlier persistent connectives ¬, ∧, as well as identity and a quantifier ∃.

> *Proposition:* Locally in each universe, Q_p is definable in this persistent predicate logic.

> *Proof:* The defining formula is a disjunction describing the top positions x, y of + triangles. For each of these, a disjunct is included of the form *there are at least x A and at least y ¬A*. ∎

Many further definability results may be found, of course.

5. In this setting, *determiners* would be binary quantifiers of type $((e, t), ((e, t), t))$, acting on situations

$$B^-$$
$$B^*$$
$$B^+$$

$$A^+ \ A^* \ A^-$$

One natural additional constraint here would be the earlier *Conservativity*, i.e., strong equivalence between QAB and $QA(B \cap A)$. Here, "$B \cap A$" stands for the partial predicate $(B^+ \cap A^+, B^- \cup A^-)$. Note that, of the obvious predicate-logical definitions for binary quantifiers in the Square of Opposition, the *universal* one lacks Conservativity as it stands:

$$\exists x(Ax \wedge Bx) \leftrightarrow \exists x(Ax \wedge (Bx \wedge Ax)),$$
$$\exists x(Ax \wedge \neg Bx) \leftrightarrow \exists x(Ax \wedge \neg(Bx \wedge Ax)),$$
$$\neg\exists x(Ax \wedge Bx) \leftrightarrow \neg\exists x(Ax \wedge (Bx \wedge Ax));$$

but $\forall x(\neg Ax \vee Bx)$ is weaker than its (conservative) counterpart

$$\forall x(\neg Ax \vee (Bx \wedge Ax)).$$

6. It would be of interest to classify various forms of *persistence* and *monotonicity* here, allowing both growth of individuals and information about them.

On the other hand, standard quantifiers also admit of interesting non-persistent variants. E.g., *all A are B* could also stand for merely $A^+ \subseteq B^+$ (or perhaps $A^+ \subseteq B^+$ & $B^- \subseteq A^-$). As so often in partial logic, our task will be to chart the reasonable *options*.

Bibliography

Adams, E.W. 1966. Probability and the Logic of Conditionals. In J. Hintikka and P. Suppes, eds., *Aspects of Inductive Logic*, Dordrecht: Reidel.

Adams, E.W. 1975. *The Logic of Conditionals*. Dordrecht: Reidel.

Allen, J. 1983. Maintaining Knowledge about Temporal Intervals. *Communications of the Association for Computing Machinery* 26, 832–843.

Almog, J., J. Perry, H. Wettstein, eds. Forthcoming. *Themes from Kaplan*. Oxford: Oxford University Press.

Aristotle. *De Interpretatione*. In J.L. Ackrill, trans., *Aristotle's Categories and De Interpretatione*, 1963, Clarandon Aristotle Series, Oxford: Clarendon Press.

Barwise, J. 1988. The Situation in Logic IV: On the Model Theory of Common Knowledge. Report CSLI–88–122, Stanford: CSLI.

Barwise, J. 1988. Three Views of Common Knowledge. In J. Halpern, ed., *Theoretical Aspects of Reasoning About Knowledge, II*, Los Altos: Morgan Kaufmann.

Barwise, J. 1987. Noun Phrases, Generalized Quantifiers and Anaphora. In P. Gärdenfors, ed., *Generalized Quantifiers: Linguistic and Logical Approaches*. Dordrecht: Reidel, 1–29.

Barwise, J., and J. Perry. 1983. *Situations and Attitudes*. Cambridge, Mass.: Bradford Books/MIT Press.

Barwise, J., and R. Cooper. 1981. Generalized Quantifiers and Natural Language. *Linguistics and Philosophy* 4, 159–219.

Bergstra, J., and J. Klop. 1984. Process Algebra for Synchronous Communication. *Information and Control* 60:1/3, 109–137.

Blamey, S. 1986. Partial Logic. In D. Gabbay and F. Guenthner, eds., *Handbook of Philosophical Logic vol. III*, Dordrecht: Reidel, 1–70.

Boolos, G. 1979. *The Unprovability of Consistency*. Cambridge: Cambridge University Press.

Bressan, A. 1972. *A General Interpreted Modal Calculus*. New Haven, Conn.: Yale University Press.

Bull, R.A. 1983, 84. Reviews in the *Journal of Symbolic Logic*, vol. 47:2 (1982), 440–445 and vol. 48:2 (1983), 488–495.

Burgess, J. 1979. Logic and Time. *Journal of Symbolic Logic*: 44, 566–582.

Burgess, J. 1981. Quick Completeness Proofs for Some Logics of Conditionals. *Notre Dame Journal of Formal Logic* 22, 76–84.

Burgess, J. 1982. Axioms for Tense Logics II: Time Periods. *Notre Dame Journal of Formal Logic* 23, 375–383.

Carnap, R. 1947. *Meaning and Necessity*. Chicago: University of Chicago Press.

Chellas, B. 1980. *Modal Logic: An Introduction*. Cambridge: Cambridge University Press.

Danecki, R. 1985. Nondeterministic Propositional Dynamic Logic with Intersection is Decidable. Lecture Notes in Computer Science 208, Berlin: Springer, 34–53.

Dowty, D. 1979. *Word Meaning and Montague Grammar*. Dordrecht: Reidel.

Feferman, S. 1984. Towards Useful Type-Free Theories I. *Journal of Symbolic Logic* 49, 75–111.

Fenstad, J.-E., P.-K. Halvorsen, T. Langholm and J. van Benthem. 1987. *Situations, Language and Logic*. Dordrecht: Reidel.

Fine, K. 1975. Vagueness, Truth and Logic. *Synthese* 30, 265–300.

Fine, K. 1977. Propositions, Sets and Properties. *Journal of Philosophical Logic* 6, 135–191.

Fine, K. 1978, 81. Model Theory for Modal Logic—Part I-III. Part I & II in *Journal of Philosophical Logic* 7, 1978, 125–156, 277–306; Part III in *Journal of Philosophical Logic* 10, 1981, 293–307.

Fine, K. 1980 & 1981. First-Order Modal Theories—I Sets, II Propositions, III Facts. Part I in *Noûs* 15, 1981, 177–205; Part II in *Studia Logica* 39, 1980, 159–202; Part III in *Synthese*, 1981, 43–122.

Fitting, M. 1969. *Intuitionistic Logic, Model Theory and Forcing.* Amsterdam: North-Holland.

Gärdenfors, P. 1981. An Epistemic Approach to Conditionals. *American Philosophical Quarterly* 68, 203–211.

Gabbay, D. 1976. *Investigations in Modal and Tense Logics.* Dordrecht: Reidel.

Gallin, D. 1975. *Intensional and Higher-Order Modal Logic.* Amsterdam: North-Holland.

Gazdar, G., et multi alii. 1987. Category Structures. Report CSLI–87–102, Stanford: CSLI.

Gödel, K. 1933. Eine Interpretation des Intuitionistischen Aussagenkalküls, *Ergebnisse eines Mathematischen Kolloquiums* 4, 39–40.

Goldblatt, R. 1971. Semantic Analysis of Orthologic. *Journal of Philosophical Logic* 1, 91–107.

Goldblatt, R. 1979. *Topoi.* Amsterdam: North-Holland.

Goldblatt, R. 1980. Diodorean Modality in Minkowski Space-Time. *Studia Logica* 39, 219–236.

Goldblatt, R. 1982. *Axiomatizing the Logic of Computer Programming.* Lecture Notes in Computer Science 130, Berlin: Springer.

Goldblatt, R. 1986. *Logics of Time and Computation.* CSLI Lecture Notes No. 7, Center for the Study of Language and Information, Stanford.

Goodman, N. 1947. The Problem of Counterfactual Conditionals. *Journal of Philosophy* 44, 113–128. Reprinted in his *Fact, Fiction and Forecast,* 1955, Cambridge, Mass.: Harvard University Press.

Gries, D. 1981. *The Science of Programming.* Berlin: Springer.

Groenendijk, J., and M. Stokhof. 1987. Dynamic Predicate Logic. Institute for Language, Logic and Information, University of Amsterdam.

Halpern, J., and Y. Moses 1984. Knowledge and Common Knowledge in a Distributed Environment. In *Proceedings of the Third ACM Conference on Principles of Distributed Computing,* 50–61.

Halpern, J., and Y. Moses. 1984. Towards a Theory of Knowledge and Ignorance. In *Proceedings of the Workshop on Non-Monotonic Reasoning,* Menlo Park, Calif.:AAAI.

Halpern, J., and Y. Shoham. 1986. A Propositional Modal Logic of Time Intervals. In *Proceedings Symposium on Logic in Computer Science,* Boston: IEEE.

Halpern, J., ed. 1986. *Theoretical Aspects of Reasoning about Knowledge*. Los Altos: Morgan Kaufmann.

Harel, D. 1984. Dynamic Logic. In D. Gabbay and F. Guenthner, eds., *Handbook of Philosophical Logic*, vol. II, Dordrecht: Reidel, 497–604.

Harper, W., et al., eds. 1981. *Ifs*. Dordrecht: Reidel.

Hayes, P. 1979. The Naive Physics Manifesto. In D. Mitchie, ed., *Expert Systems*, Edinburgh University Press. Also in J. Hobbs, ed., 1985, *Commonsense Summer: Final Report*, Report No. CSLI–85–35, Stanford: CSLI.

Hintikka, J. 1962. *Knowledge and Belief: An Introduction to the Logic of the Two Notions*. Ithaca, N.Y.: Cornell University Press.

Hintikka, J. 1969. *Models for Modalities*. Dordrecht: Reidel.

Hintikka, J. 1975. *The Intentions of Intentionality and Other New Models for the Modalities*. Dordrecht: Reidel.

Hughes, G., and M. Cresswell. 1968. *An Introduction to Modal Logic*. London: Methuen.

Hughes, G., and M. Creswell. 1984. *A Companion to Modal Logic*. London: Methuen.

Humberstone, L. 1979. Interval Semantics for Tense Logics, *Journal of Philosophical Logic* 8, 171–196.

Humberstone, L. 1981. From Worlds to Possibilities. *Journal of Philosophical Logic* 10, 313–344.

Kamp, H. 1979. Instants, Events and Temporal Discourse. In R. Bäuerle, et al., eds., *Semantics From Different Points of View*. Berlin: Springer, 376–417.

Kamp, H. 1960. *Tense Logic and the Theory of Linear Order*, Ph.D. dissertation, Department of Philosophy, University of California, Los Angeles.

Kamp, H. 1971. Formal Properties of 'Now'. *Theoria* 37, 227–273.

Kamp, H. 1981. A Theory of Truth and Semantic Interpretation. In J. Groenendijk, T. Janssen, and M. Stokhof, eds., *Formal Methods in the Study of Language*, volume I, Mathematische Centrum, Amsterdam. Reprinted in J. Groenendijk, T. Janssen, and M. Stokhof, eds., *Truth, Interpretation and Information*, 1984, Dordrecht: Foris.

Kamp, H. 1984. A Scenic Tour Through the Land of Naked Infinitives. To appear in *Linguistics and Philosophy*.

Kasper, B., and W. Rounds. 1987. The Logic of Unification in Grammar. To appear in *Linguistics and Philosophy*.

Keenan, E., and L. Faltz. 1985. *Boolean Semantics for Natural Language*. Dordrecht: Reidel.

Keenan. E., and Y. Stavi. 1986. A Semantic Characterization of Natural Language Determiners. *Linguistics and Philosophy* 9, 253–326.

Kratzer, A. 1981. The Notional Category of Modality. In H.-J. Eikmeyer, et al., eds., *Words, Worlds and Contexts*. Berlin: W. de Gruyter.

Kratzer, A. 1979. Conditional Necessity and Possibility. In R. Bäuerle, U. Egli and A. von Stechow, eds., *Semantics from Different Points of View*. Berlin: Springer.

Kratzer, A. 1976. Partition and Revision: The Semantics of Counterfactuals. *Journal of Philosophical Logic*, 201–216.

Kripke, S. 1980. *Naming and Necessity*. Cambridge, Mass.: Harvard University Press.

Kripke, S. 1959. A Completeness Theorem in Modal Logic. *Journal of Symbolic Logic* 24, 1–14.

Kripke, S. 1963. Semantical Considerations on Modal Logic. *Acta Philosophica Fennica* 16, 83–94. Reprinted in L. Linsky, ed., *Reference and Modality*, 1971, Oxford: Oxford University Press.

Kripke, S. 1965. Semantical Analysis of Intuitionistic Logic I. In J.N. Crossley and M. Dummett, eds., *Formal Systems and Recursive Functions*, Amsterdam: North Holland.

Kripke, S. 1971. Identity and Necessity. In M. Munitz, ed., *Identity and Individuation*, New York: New York University Press, 135–164.

Ladkin, P. 1987. Models of Axioms for Time Intervals. Palo Alto, Calif.: Kestrel Institute.

Ladkin, P., and R. Maddux. 1987. The Algebra of Convex Time Intervals. Kestrel Institute, Palo Alto/Department of Mathematics, Iowa State University at Ames.

Lamport, L. 1985. *Interprocess Communication. Final Report.* SRI International, Menlo Park.

Landman, F. 1986. *Towards a Theory of Information. The Status of Partial Objects in Semantics.* GRASS series Vol. 6, Dordrecht: Foris.

Langholm, T. 1984. Some Tentative Systems Relating to Situation Semantics. In *Report of an Oslo Seminar in Logic and Linguistics*, Preprint series 9, Mathematical Institute, University of Oslo.

Langholm, T. 1987. *Partiality, Truth and Persistence.* Ph.D. dissertation, Department of Philosophy, Stanford University. Forthcoming CSLI Lecture Notes.

Langholm, T. 1988. H. B. Smith on Modality: A Logical Reconstruction. *Journal of Philosophical Logic* 16:4, 337—346.

Lenzen, W. 1980. *Glauben, Wissen und Wahrscheinlichkeit.* Vienna: Springer.

Lewis, C. I. 1910. *A Survey of Symbolic Logic,* Berkeley, Calif.: University of California Press.

Lewis, D. 1968. Counterpart Theory for Quantified Modal Logic. *Journal of Philosophy* 65, 113–26. Reprinted in J. Loux *The Possible and the Actual,* 1979, Ithaca, N.Y.: Cornell University Press, 110–128; and, with postscripts, in *Philosophical Papers,* volume I, 26–46.

Lewis, D. 1969. *Convention. A Philosophical Study.* Cambridge: Harvard University Press.

Lewis, D. 1973. Counterfactuals and Comparative Possibility. *Journal of Philosophical Logic* 2, 418–446. Reprinted in *Philosophical Papers,* volume II, 3–31.

Lewis, D. 1981. Ordering Semantics and Premise Semantics for Counterfactuals. *Journal of Philosophical Logic* 10, 217–34.

Lewis, D. 1983, 86. *Philosophical Papers,* volumes I & II. Oxford: Oxford University Press.

Lewis, D. 1986. *On the Plurality of Worlds.* Oxford: Basil Blackwell.

Lukasiewiz., J. 1920. On 3-valued Logic. In S. McCall, ed., *Polish Logic,* 1930, Oxford: Oxford University Press.

Lukasiewiz., J. 1930. Many-valued Systems of Propositional Logic. In S. McCall, ed., *Polish Logic,* 1930, Oxford: Oxford University Press.

Lifschitz, V. 1986. Computing Circumscription. Department of Computer Science, Stanford University.

McCarthy, J. 1980. Circumscription—A Form of Non-Monotonic Reasoning. *Artificial Intelligence* 13, 295–323.

McTaggart, J. E. 1908. The Unreality of Time. *Mind* 18, 457–84. In *The Nature of Existence,* vol. 2, 1927, Cambridge: Cambridge University Press.

Meyer, J.-J. 1984. Deontic Logic Viewed as a Variant of Dynamic Logic. Report lR-93, Department of Mathematics and Computer Science, Free University, Amsterdam.

Montague, R. 1974. *Formal Philosophy,* R. Thomason, ed. New Haven, Conn.: Yale University Press.

Moore, R. 1983. *Semantical Considerations on Non-Monotonic Logic.* SRI International, Menlo Park.

Moore, R. 1984. *A Formal Theory of Knowledge and Action.* Technical Note 320, Artificial Intelligence Center, SRI International, Menlo Park.

Muskens, R. 1987. *Going Partial.* Institute for Language, Logic and Information, University of Amsterdam. In *Montague Grammar.*

Muskens, R. 1987. *Going Relational in Type Theory.* Institute for Language, Logic and Information, University of Amsterdam.

Nute, D. 1986. *A Non-Monotonic Logic based on Conditional Logic.* Report 01–0007, Advanced Computational Methods Center, University of Georgia, Athens, Georgia.

Orlowska, E. 1985. Logic of Indiscernibility Relations. Lecture Notes in Computer Science 208, Berlin: Springer, 177–186.

Perry, J. 1986. From Worlds to Situations. *Journal of Philosophical Logic* 15:1, 83–107.

Plantinga, A. 1974. *The Nature of Necessity.* Oxford: Clarendon Press.

Pnueli, A. 1981. The Temporal Semantics of Concurrent Programs. *Theoretical Computer Science* 13, 45–60.

Pratt, V. 1980. Applications of Modal Logic to Programming. *Studia Logica* 39, 257–274.

Prior, A. 1967. *Past, Present and Future.* Oxford: Oxford Clarendon Press.

Prior, A., and K. Fine. 1977. *Worlds, Times and Selves.* Amherst: The University of Massachusetts Press.

Quine, W. V. O. 1953. *From a Logical Point of View.* Cambridge, Mass.: Harvard University Press.

Quine, W. V. O. 1953. Reference and Modality. In Quine, 1953. Reprinted in L. Linsky, ed., *Reference and Modality*, 1971, Oxford: Oxford University Press.

Quine, W. V. O. 1960. *Word and Object.* Cambridge: MIT Press.

Ramsey, F. P. 1950. *The Foundations of Mathematics and Other Logical Essays*, R.B. Braithwaithe, ed. London: Routledge & Kegan Paul.

Rescher, N. 1961. Belief-Contravening Suppositions and the Problem of Contrary-to-Fact Conditionals. *The Philosophical Review* 60, 176–196. Reprinted in E. Sosa, ed., *Causation and Conditionals*, 1975, Oxford: Oxford University Press, 156–164.

Röper, P. 1980. Intervals and Tenses. *Journal of Philosophical Logic* 9, 451–469.

Rooth, M., and B. Partee. 1983. Generalized Conjunction and Type Ambiguity. In R. Bäuerle et al., eds., *Meaning, Use and Interpretation of Language*, Berlin: De Gruyter, 361–383.

Rosenbloom, P. 1950. *The Elements of Mathematical Logic*, New York: Dover.

Rosenschein, S., and L. Kaebling. 1987. The Synthesis of Digital Machines with Provable Epistemic Properties. Technical Note 412, SRI International, Menlo Park, CA.

Russell, B. 1926. *Our Knowledge of the External World*, London: Allen & Unwin.

Salmon, W. 1977. Laws, Modalities and Counterfactuals. *Synthese* 35, 191–229.

Shapiro, S., ed. 1985 *Intensional Mathematics*, Amsterdam: North-Holland.

Shehtman, V. B. 1983. Modal Logics of Domains on the Real Plane, *Studia Logica* 42:1, 63–80.

Shieber, S. 1986. *An Introduction to Unification-Based Approaches to Grammar.* CSLI Lecture Notes No. 4, Center for the Study of Language and Information, Stanford University.

Shoham, Y. 1986. *Reasoning about Change: Time and Causation from the Standpoint of Artificial Intelligence.* Cambridge, Mass.: The MIT Press.

Sosa, E., ed. 1975. *Causation and Conditionals.* Oxford: Oxford University Press.

Stalnaker, R. 1968. A Theory of Conditionals. In N. Rescher, ed., *Studies in Logical Theory*, Oxford: Basil Blackwell, 98–112. Reprinted in E. Sosa, ed., *Causation and Conditionals*, Oxford: Oxford University Press, 165–179.

Stalnaker, R. 1970. Probability and Conditionals. *Philosophy of Science* 37, 68–80.

Stalnaker, R. 1972. Pragmatics. In D. Davidson and G. Harman, eds., *Semantics of Natural Language*, Dordrecht: Reidel, 380–397.

Stalnaker, R. 1984. *Inquiry.* Cambridge, Mass.: The MIT Press.

Stalnaker, R. 1986. Possible Worlds and Situations. *Journal of Philosophical Logic* 15:1, 109–123.

Stalnaker, R., and R. Thomason. 1970. A Semantic Analysis of Conditional Logic. *Theoria* 36, 23–42.

Taylor, R. 1962. Fatalism. *The Philosophical Review* 71, 56–66.

Thijsse, E. 1987. *Kripke Models for Knowledge Bases I: S5-Miniatures.* Department of Language and Informatics, University of Tilburg.

Thomason, R. 1984. Combinations of Tense and Modality. In D. Gabbay and F. Guenther, eds., *Handbook of Philosophical Logic*, vol. II, Dordrecht: Reidel, 135–165.

Thomason, S. K. 1979. *Possible Worlds, Times and Tenure.* Department of Mathematics, Simon Fraser University, Vancouver.

Troelstra, A., and D. van Dalen. 1988. *Principles of Constructive Mathematics*, vol. I and II, Amsterdam: North-Holland.

Turner, R. 1981. Counterfactuals Without Possible Worlds. *Journal of Philosophical Logic* 10, 453–493.

Urquhart, A. 1986. Many-Valued Logic. In D. Gabbay and F. Guenthner, eds., *Handbook of Philosphical Logic*, vol. III, Dordrecht: Reidel, 71–116.

Vakarelov, D. 1987. *S4* and *S5* together—*S4+5*. Sector of Mathematical Logic, University of Sofia.

Van Benthem, J. 1981. Possible Worlds Semantics for Classical Logic. Report ZW–8018, Department of Mathematics, Rijksuniversiteit, Groningen.

Van Benthem, J. 1983. Determiners and Logic. *Linguistics and Philosophy* 6, 447–478,

Van Benthem, J. 1983. *The Logic of Time.* Dordrecht: Reidel.

Van Benthem, J. 1984. Correspondence Theory. In D. Gabbay and F. Guenthner, eds., *Handbook of Philosophical Logic*, vol. II. Dordrecht: Reidel, 167–247.

Van Benthem, J. 1984. Foundations of Conditional Logic. *Journal of Philosophical Logic* 13:3, 303–349.

Van Benthem, J. 1984. Possible Worlds Semantics: a research program that cannot fail? *Studia Logica* 43:4, 379–393.

Van Benthem, J. 1984. Tense, Logic and Time. *Notre Dame Journal of Formal Logic* 25, 1–16.

Van Benthem, J. 1984. *Minimal Conditions for Interval Models.* CSLI, Stanford University.

Van Benthem, J. 1985. *Modal Logic and Classical Logic.* Napoli, Atlantic Heights: Bibliopolis/The Humanities Press.

Van Benthem, J. 1986. A Linguistic Turn: New Directions in Logic. In R. Marcus et al., eds., *Proceedings 7th International Congress of Logic, Methodology and Philosophy of Science, Salzburg 1983*, Amsterdam: North-Holland, 205–240.

Van Benthem, J. 1986. Partiality and Nonmonotonicity in Classical Logic. *Logique et Analyse* 29, 225–247.

Van Benthem, J. 1986. Tenses in Real Time. *Zeitschrift für mathematische Logik und Grundlagen der Mathematik* 32, 61–72.

Van Benthem, J. 1986. *Essays in Logical Semantics.* Dordrecht: Reidel.

Van Benthem, J. 1987. Verisimilitude and Conditionals. In T. Kuipers, ed., *What is Closer-to-the-Truth?*, Amsterdam: Rodopi, 103–128.

Van Benthem, J. 1987. *Semantics of Programming Languages.* Faculteit Wiskunde en Informatica, University of Amsterdam.

Van Benthem, J. 1987. Categorial Grammar and Type Theory. Report 87–07, Institute for Language, Logic and Information, University of Amsterdam. To appear in *Linguistics and Philosophy.*

Van Benthem, J. 1988. Parallels in the Semantics of Natural Languages and Programming Languages. In M. Garrido, ed., *Logic Colloquium. Granada 1987*, Amsterdam: North-Holland.

Van Benthem, J., and J. van Eyck. 1982. The Dynamics of Interpretation. *Journal of Semantics* 1, 3–20.

Van Eck, J. 1981. *A System of Temporally Relative Modal and Deontic Logic, with Philosophical Applications.* Ph.D. dissertation, Rijksuniversiteit, Groningen. Also in *Logique et Analyse*, 1983.

Veltman, F. 1981. Data Semantics. In J. Groenendijk et al., eds., *Formal Methods in the Study of Language*, Mathematical Centre, Amsterdam. Reprinted in the GRASS-series, Vol. II, 1984, Dordrecht: Foris.

Veltman, F. 1976. Prejudices, Presuppositions and the Theory of Conditionals. In J. Groenendijk and M. Stokhof, eds., *Amsterdam Papers in Formal Grammar*, volume I, Centrale Interfaculteit, University of Amsterdam.

Veltman, F. 1985. *Logics for Conditionals.* Ph.D. dissertation, Filosofisch Instituut, University of Amsterdam. To appear Cambridge: Cambridge University Press.

Von Wright, G.H. 1983 ff. *Philosophical Papers*, volumes I-III. Volume I, Practical Reason (1983); volume II, Philosophical Logic; volume III, Proof, Knowledge, and Modality. Ithaca: Cornell University Press.

White, M. 1984. The Necessity of the Past and Modal-Tense Logic Incompleteness. *Notre Dame Journal of Formal Logic*, 25:1, 59–71.

Winnie, J. 1977. The Causal Theory of Space-Time. In J. Earman, et al., eds., *Foundations of Space-Time Theories*, University of Minnesota Press, Minneapolis, 134–205.

Zwarts, F. 1986. *Categorial Grammar and Algebraic Semantics.* Ph.D. dissertation, Nederlands Instituut, Rijksuniversiteit, Groningen. To appear Dordrecht and Boston: Reidel.

Index

CSLI Publications

Reports

The following titles have been published in the CSLI Reports series. These reports may be obtained from CSLI Publications, Ventura Hall, Stanford University, Stanford, CA 94305-4115.

The Situation in Logic–I Jon Barwise CSLI–84–2 (*$2.00*)

Coordination and How to Distinguish Categories Ivan Sag, Gerald Gazdar, Thomas Wasow, and Steven Weisler CSLI–84–3 (*$3.50*)

Belief and Incompleteness Kurt Konolige CSLI–84–4 (*$4.50*)

Equality, Types, Modules and Generics for Logic Programming Joseph Goguen and José Meseguer CSLI–84–5 (*$2.50*)

Lessons from Bolzano Johan van Benthem CSLI–84–6 (*$1.50*)

Self-propagating Search: A Unified Theory of Memory Pentti Kanerva CSLI–84–7 (*$9.00*)

Reflection and Semantics in LISP Brian Cantwell Smith CSLI–84–8 (*$2.50*)

The Implementation of Procedurally Reflective Languages Jim des Rivières and Brian Cantwell Smith CSLI–84–9 (*$3.00*)

Parameterized Programming Joseph Goguen CSLI–84–10 (*$3.50*)

Morphological Constraints on Scandinavian Tone Accent Meg Withgott and Per-Kristian Halvorsen CSLI–84–11 (*$2.50*)

Partiality and Nonmonotonicity in Classical Logic Johan van Benthem CSLI–84–12 (*$2.00*)

Shifting Situations and Shaken Attitudes Jon Barwise and John Perry CSLI–84–13 (*$4.50*)

Aspectual Classes in Situation Semantics Robin Cooper CSLI–85–14–C (*$4.00*)

Completeness of Many-Sorted Equational Logic Joseph Goguen and José Meseguer CSLI–84–15 (*$2.50*)

Moving the Semantic Fulcrum Terry Winograd CSLI–84–17 (*$1.50*)

On the Mathematical Properties of Linguistic Theories C. Raymond Perrault CSLI–84–18 (*$3.00*)

A Simple and Efficient Implementation of Higher-order Functions in LISP Michael P. Georgeff and Stephen F.Bodnar CSLI–84–19 (*$4.50*)

On the Axiomatization of "if-then-else" Irène Guessarian and José Meseguer CSLI–85–20 (*$3.00*)

The Situation in Logic–II: Conditionals and Conditional Information Jon Barwise CSLI–84–21 (*$3.00*)

Principles of OBJ2 Kokichi Futatsugi, Joseph A. Goguen, Jean-Pierre Jouannaud, and José Meseguer CSLI–85–22 (*$2.00*)

Querying Logical Databases Moshe Vardi CSLI–85–23 (*$1.50*)

Computationally Relevant Properties of Natural Languages and Their Grammar Gerald Gazdar and Geoff Pullum CSLI–85–24 (*$3.50*)

An Internal Semantics for Modal Logic: Preliminary Report Ronald Fagin and Moshe Vardi CSLI–85–25 (*$2.00*)

The Situation in Logic–III: Situations, Sets and the Axiom of Foundation Jon Barwise CSLI–85–26 (*$2.50*)

Semantic Automata Johan van Benthem CSLI–85–27 (*$2.50*)

Restrictive and Non-Restrictive Modification Peter Sells CSLI–85–28 (*$3.00*)

Institutions: Abstract Model Theory for Computer Science J. A. Goguen and R. M. Burstall CSLI–85–30 (*$4.50*)

A Formal Theory of Knowledge and Action Robert C. Moore CSLI–85–31 ($5.50)

Finite State Morphology: A Review of Koskenniemi (1983) Gerald Gazdar CSLI–85–32 ($1.50)

The Role of Logic in Artificial Intelligence Robert C. Moore CSLI–85–33 ($2.00)

Applicability of Indexed Grammars to Natural Languages Gerald Gazdar CSLI–85–34 ($2.00)

Commonsense Summer: Final Report Jerry R. Hobbs, et al CSLI–85–35 ($12.00)

Limits of Correctness in Computers Brian Cantwell Smith CSLI–85–36 ($2.50)

On the Coherence and Structure of Discourse Jerry R. Hobbs CSLI–85–37 ($3.00)

The Coherence of Incoherent Discourse Jerry R. Hobbs and Michael H. Agar CSLI–85–38 ($2.50)

The Structures of Discourse Structure Barbara Grosz and Candace L. Sidner CSLI–85–39 ($4.50)

A Complete, Type-free "Second-order" Logic and Its Philosophical Foundations Christopher Menzel CSLI–86–40 ($4.50)

Possible-world Semantics for Autoepistemic Logic Robert C. Moore CSLI–85–41 ($2.00)

Deduction with Many-Sorted Rewrite José Meseguer and Joseph A. Goguen CSLI–85–42 ($1.50)

On Some Formal Properties of Metarules Hans Uszkoreit and Stanley Peters CSLI–85–43 ($1.50)

Language, Mind, and Information John Perry CSLI–85–44 ($2.00)

Constraints on Order Hans Uszkoreit CSLI–86–46 ($3.00)

Linear Precedence in Discontinuous Constituents: Complex Fronting in German Hans Uszkoreit CSLI–86–47 ($2.50)

A Compilation of Papers on Unification-Based Grammar Formalisms, Parts I and II Stuart M. Shieber, Fernando C.N. Pereira, Lauri Karttunen, and Martin Kay CSLI–86–48 ($10.00)

An Algorithm for Generating Quantifier Scopings Jerry R. Hobbs and Stuart M. Shieber CSLI–86–49 ($2.50)

Verbs of Change, Causation, and Time Dorit Abusch CSLI–86–50 ($2.00)

Noun-Phrase Interpretation Mats Rooth CSLI–86–51 ($2.00)

Noun Phrases, Generalized Quantifiers and Anaphora Jon Barwise CSLI–86–52 ($2.50)

Circumstantial Attitudes and Benevolent Cognition John Perry CSLI–86–53 ($1.50)

A Study in the Foundations of Programming Methodology: Specifications, Institutions, Charters and Parchments Joseph A. Goguen and R. M. Burstall CSLI–86–54 ($2.50)

Quantifiers in Formal and Natural Languages Dag Westerståhl CSLI–86–55 ($7.50)

Intentionality, Information, and Matter Ivan Blair CSLI–86–56 ($3.00)

Graphs and Grammars William Marsh CSLI–86–57 ($2.00)

Computer Aids for Comparative Dictionaries Mark Johnson CSLI–86–58 ($2.00)

The Relevance of Computational Linguistics Lauri Karttunen CSLI–86–59 ($2.50)

Grammatical Hierarchy and Linear Precedence Ivan A. Sag CSLI–86–60 ($3.50)

D-PATR: A Development Environment for Unification-Based Grammars Lauri Karttunen CSLI–86–61 ($4.00)

A Sheaf-Theoretic Model of Concurrency Luís F. Monteiro and Fernando C. N. Pereira CSLI–86–62 ($3.00)

Discourse, Anaphora and Parsing Mark Johnson and Ewan Klein CSLI–86–63 ($2.00)

Tarski on Truth and Logical Consequence John Etchemendy CSLI–86–64 (*$3.50*)

The LFG Treatment of Discontinuity and the Double Infinitive Construction in Dutch Mark Johnson CSLI–86–65 (*$2.50*)

Categorial Unification Grammars Hans Uszkoreit CSLI–86–66 (*$2.50*)

Generalized Quantifiers and Plurals Godehard Link CSLI–86–67 (*$2.00*)

Radical Lexicalism Lauri Karttunen CSLI–86–68 (*$2.50*)

Understanding Computers and Cognition: Four Reviews and a Response Mark Stefik, Editor CSLI–87–70 (*$3.50*)

The Correspondence Continuum Brian Cantwell Smith CSLI–87–71 (*$4.00*)

The Role of Propositional Objects of Belief in Action David J. Israel CSLI–87–72 (*$2.50*)

From Worlds to Situations John Perry CSLI–87–73 (*$2.00*)

Two Replies Jon Barwise CSLI–87–74 (*$3.00*)

Semantics of Clocks Brian Cantwell Smith CSLI–87–75 (*$2.50*)

Varieties of Self-Reference Brian Cantwell Smith CSLI–87–76 (*Forthcoming*)

The Parts of Perception Alexander Pentland CSLI–87–77 (*$4.00*)

Topic, Pronoun, and Agreement in Chicheŵa Joan Bresnan and S. A. Mchombo CSLI–87–78 (*$5.00*)

HPSG: An Informal Synopsis Carl Pollard and Ivan A. Sag CSLI–87–79 (*$4.50*)

The Situated Processing of Situated Language Susan Stucky CSLI–87–80 (*Forthcoming*)

Muir: A Tool for Language Design Terry Winograd CSLI–87–81 (*$2.50*)

Final Algebras, Cosemicomputable Algebras, and Degrees of Unsolvability Lawrence S. Moss, José Meseguer, and Joseph A. Goguen CSLI–87–82 (*$3.00*)

The Synthesis of Digital Machines with Provable Epistemic Properties Stanley J. Rosenschein and Leslie Pack Kaelbling CSLI–87–83 (*$3.50*)

Formal Theories of Knowledge in AI and Robotics Stanley J. Rosenschein CSLI–87–84 (*$1.50*)

An Architecture for Intelligent Reactive Systems Leslie Pack Kaelbling CSLI–87–85 (*$2.00*)

Order-Sorted Unification José Meseguer, Joseph A. Goguen, and Gert Smolka CSLI–87–86 (*$2.50*)

Modular Algebraic Specification of Some Basic Geometrical Constructions Joseph A. Goguen CSLI–87–87 (*$2.50*)

Persistence, Intention and Commitment Phil Cohen and Hector Levesque CSLI–87–88 (*$3.50*)

Rational Interaction as the Basis for Communication Phil Cohen and Hector Levesque CSLI–87–89 (*Forthcoming*)

An Application of Default Logic to Speech Act Theory C. Raymond Perrault CSLI–87–90 (*$2.50*)

Models and Equality for Logical Programming Joseph A. Goguen and José Meseguer CSLI–87–91 (*$3.00*)

Order-Sorted Algebra Solves the Constructor-Selector, Mulitple Representation and Coercion Problems Joseph A. Goguen and José Meseguer CSLI–87–92 (*$2.00*)

Extensions and Foundations for Object-Oriented Programming Joseph A. Goguen and José Meseguer CSLI–87–93 (*$3.50*)

L3 Reference Manual: Version 2.19 William Poser CSLI–87–94 (*$2.50*)

Change, Process and Events Carol E. Cleland CSLI–87–95 (*Forthcoming*)

One, None, a Hundred Thousand Specification Languages Joseph A. Goguen CSLI–87–96 (*$2.00*)

Constituent Coordination in HPSG Derek Proudian and David Goddeau CSLI–87–97 (*$1.50*)

Lecture Notes

The titles in this series are distributed by the University of Chicago Press and may be purchased in academic or university bookstores or ordered directly from the distributor at 5801 Ellis Avenue, Chicago, Illinois 60637.

A Manual of Intensional Logic Johan van Benthem, second edition. Lecture Notes No. 1

Emotions and Focus Helen Fay Nissenbaum. Lecture Notes No. 2

Lectures on Contemporary Syntactic Theories Peter Sells. Lecture Notes No. 3

An Introduction to Unification-Based Approaches to Grammar Stuart M. Shieber. Lecture Notes No. 4

The Semantics of Destructive Lisp Ian A. Mason. Lecture Notes No. 5

An Essay on Facts Ken Olson. Lecture Notes No. 6

Logics of Time and Computation Robert Goldblatt. Lecture Notes No. 7

Word Order and Constituent Structure in German Hans Uszkoreit. Lecture Notes No. 8

Color and Color Perception: A Study in Anthropocentric Realism David Russel Hilbert. Lecture Notes No. 9

Prolog and Natural-Language Analysis Fernando C. N. Pereira and Stuart M. Shieber. Lecture Notes No. 10

Working Papers in Grammatical Theory and Discourse Structure: Interactions of Morphology, Syntax, and Discourse M. Iida, S. Wechsler, and D. Zec (Eds.) with an Introduction by Joan Bresnan. Lecture Notes No. 11

Natural Language Processing in the 1980s: A Bibliography Gerald Gazdar, Alex Franz, Karen Osborne, and Roger Evans. Lecture Notes No. 12

Information-Based Syntax and Semantics Carl Pollard and Ivan Sag. Lecture Notes No. 13

Non-Well-Founded Sets Peter Aczel. Lecture Notes No. 14

Partiality, Truth and Persistence Tore Langholm. Lecture Notes No. 15

A Logic of Attribute-Value Structures and the Theory of Grammar Mark Johnson.